STRUCTURE
AND BONDING

Volume 35

Editors:
J. D. Dunitz, Zürich · J. B. Goodenough, Oxford
P. Hemmerich, Konstanz · J. A. Ibers, Evanston
C. K. Jørgensen, Genève · J. B. Neilands, Berkeley
D. Reinen, Marburg · R. J. P. Williams, Oxford

With 41 Figures and 15 Tables

Springer-Verlag Berlin Heidelberg GmbH 1978

ISBN 978-3-662-15458-8   ISBN 978-3-540-35793-3 (eBook)
DOI 10.1007/978-3-540-35793-3

Library of Congress Catalog Card Number 67-11280

© by Springer-Verlag Berlin Heidelberg 1978

Originally published by Springer-Verlag Berlin Heidelberg New York in 1978.

Softcover reprint of the hardcover 1st edition 1978

2152/3140-543210

Contents

STRUCTURE AND BONDING is issued at irregular intervals, according to the material received. With the acceptance for publication of a manuscript, copyright of all countries is vested exclusively in the publisher. Only papers not previously published elsewhere should be submitted. Likewise, the author guarantees against subsequent publication elsewhere. The text should be as clear and concise as possible, the manuscript written on one side of the paper only. Illustrations should be limited to those actually necessary.

Manuscripts will be accepted by the editors:

Professor Dr. *Jack D. Dunitz*	Laboratorium für Organische Chemie der Eidgenössischen Hochschule CH-8006 Zürich, Universitätsstraße 6/8
Professor *John B. Goodenough*	Inorganic Chemistry Laboratory University of Oxford, South Parks Road Oxford OX1 3QR, Great Britain
Professor Dr. *Peter Hemmerich*	Universität Konstanz, Fachbereich Biologie D-7750 Konstanz, Postfach 733
Professor *James A. Ibers*	Department of Chemistry, Northwestern University Evanston, Illinois 60201, U.S.A.
Professor Dr. *C. Klixbüll Jørgensen*	Dépt. de Chimie Minérale de l'Université 30 quai Ernest Ansermet, CH-1211 Genève 4
Professor *Joe B. Neilands*	University of California, Biochemistry Department Berkeley, California 94720, U.S.A.
Professor Dr. *Dirk Reinen*	Fachbereich Chemie der Universität Marburg D-3550 Marburg, Gutenbergstraße 18
Professor *Robert Joseph P. Williams*	Wadham College, Inorganic Chemistry Laboratory Oxford OX1 3QR, Great Britain

SPRINGER-VERLAG

D-6900 Heidelberg 1
P. O. Box 105280
Telephone (06221) 487·1
Telex 04-61723

D-1000 Berlin 33
Heidelberger Platz 3
Telephone (030) 822001
Telex 01-83319

SPRINGER-VERLAG
NEW YORK INC.

175, Fifth Avenue
New York, N.Y. 10010
Telephone (212) 477-8200

Conformational Analysis in Inorganic Chemistry: Semi-Empirical Quantum Calculation vs. Experiment

To EVIE

Jean-François Labarre

Laboratoire de Chimie de Coordination du CNRS, BP 4142, 31030 Toulouse Cedex, France

Table of Contents

 In order to avoid a provocative interpretation of the title of this paper (*semi-empirical* being a word that tends to make pure quantum chemists irascible and pure experimentalists suspicious), here is a little story to illustrate the author's approach:

 "One night, two theologians and a. . . Publican were discussing God and Faith. Suddenly, the fuse blew and, of course, the lights went out. The theologians continued arguing, going into the theological reasons and the more or less divine character of such a failure. The Publican, however, went to mend the fuse."

 I must say that recent criticism from pure quantum chemists of the use of semi-empirical methods for precise and well-defined chemical purposes has made me feel a bit like the Publicans. The purpose of this paper is to present our conclusions and to suggest how they may be of benefit to chemists.

I. Our Starting Point: Conformational Analysis of Lewis Adducts

Having worked for many years on chemistry and the behavior of Lewis adducts in nonmetallic coordination chemistry, we realized in 1970 that conformational analysis, both experimental and theoretical, of such compounds needed to be developed.

A literature survey showed that, among the numerous Lewis adducts whose chemical and physical behavior had been extensively studied, only one, the trifluoro-phosphine-borane complex, $F_3P \cdot BH_3$, had been experimentally investigated in the gas phase from a geometrical and conformational point of view by means of micro-wave spectroscopy (1). This complex has an "ethane-like" structure, the staggered conformation (Fig. 1) being preferred by (3.24 ± 0.15) kcal. mole^{-1} to the eclipsed one; its dipole moment as measured by the Stark effect equals (1.64 ± 0.02) D and the (P.B) bond length is 1.836 Å. Note that the bond lengths and valency angles are given with remarkable precision, despite the lack of isotopic derivatives due to the absence of stable isotopes for P and F.

Such work could be used to test the validity of a quantum approach to preferred conformations of Lewis adducts. We therefore decided to perform such a test by means of semi-empirical methods in the hope of providing chemists with *an inexpensive, rapid, reliable* tool for the study of large *series* of compounds.

Thus, starting with the geometries of the F_3P and BH_3 parts as in the complex, and assuming that these geometries remain unaltered upon rotation around the (P.B) direction (rigid rotor assumption), we used the CNDO/2 formalism within its original parametrization (2) to compute the variation of the total energy, E_{tot}, (Fig. 2) as a function of the rotation angle θ around the (P.B) bond. The staggered conformation

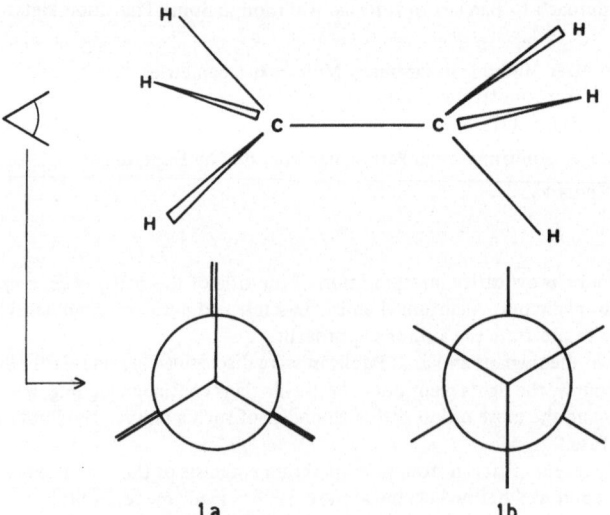

1 a   1 b

Fig. 1 a and b. Staggered and eclipsed conformations of ethane

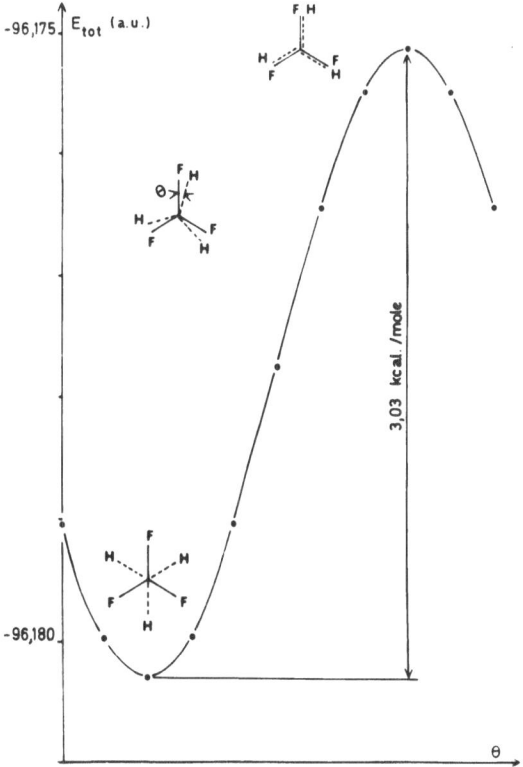

Fig. 2. Variation of the total energy of $F_3P \cdot BH_3$ vs. the rotation angle θ around the (P.B) bond

was found to be preferred with the calculated barrier to internal rotation (staggered/eclipsed) being 3.03 kcal. $mole^{-1}$, in very good agreement with the corresponding microwave data. The theoretical dipole moment for this staggered form was 1.615 D, to be compared to the experimental value of 1.64 ± 0.02. Moreover, upon plotting the variation of E_{tot} relative to an approach of the F_3P and BH_3 parts along their common C_3 axis (in the staggered form), we observed that the potential well was obtained for an optimized (P.B) bond length of 1.836 Å, i.e. exactly the microwave determination (3).

This excellent agreement encouraged us to collaborate on further work in the same field with certain experimentalists, namely *R. L. Kuczkowski* (Ann Arbor) and *H. Dreizler* (Kiel) for microwave spectroscopy, and *L. V. Vilkov* (Moscow) and *I. Hargittai* (Budapest) for electron diffraction. The CNDO/2 method proved able to reproduce accurately experimental data, mainly preferred conformations, rotational barriers, dipole moments, and other monoelectronic properties. This was the case for $F_2HP \cdot BH_3$ (3) (microwave study by *Pasinski* and *Kuczkowski* (4)), $H_3B \cdot CO$ (5), $(CH_3)H_2P \cdot BH_3$ (6) and $(CH_3)_3P \cdot BH_3$ (7) (microwave study by *Bryan* and

3

Kuczkowski (8)) and related compounds, where the agreement between theory and experiment appeared to be as good as with $F_3P \cdot BH_3$.

We concluded that a CNDO/2 approach could help us not only to *reproduce* conformational observables but also to *understand the origin* of preferred conformations and of any related parameter. Clearly, this facility must be connected with the fact that Lewis adducts or, more generally, *inorganic* compounds can be qualified as *electronically localized systems*. The existence of a clearly defined field of potential applications was simultaneously demonstrated by *Gropen* and *Seip (9)* and *Perahia* and *Pullman (10)*.

On the basis of these encouraging results, we decided to pursue two main areas:
(1) to try to understand the origin of preferred conformations, and
(2) to guide experimentalists in using their techniques to solve a geometrical or conformational structure, particularly when studying molecules containing hydrogen atoms.
The accurate spatial location of these atoms generally needs a sophisticated approach, for example, the study of a *complete* deuterated set of isotopic derivatives in microwave spectroscopy or the use of neutron diffraction techniques. We shall see below that a set of CNDO/2 calculations combined with suitable experiments (microwave spectroscopy and/or electron diffraction) may help to solve the geometrical and conformational analysis of compounds containing many hydrogen atoms.

There was another idea we wished to test in order to eliminate the necessity of knowing the accurate geometry of a molecule before being able to calculate its electron distribution and monoelectronic properties. In dealing with the *evolution of behavior* within large *series* of homologous compounds, *Gordon* and *Pople (11)* established the validity of the so-called *Standard Geometrical Model* in *organic chemistry*. They showed that calculations based on *constant* bond lengths and angles throughout a series allow the trend of any desired property to be determined. Such an assumption had never before been tested in the field of *inorganic* or *coordination chemistry*. This we did and we shall show how useful this model is.

II. The Origin of Preferred Conformations

A consideration of the analysis of rotational conformations in a molecule suggests that the existence of a preferred conformation is due to an optimized stabilizing balance between the mono- and polycentric energy terms, which vary upon rotation.

Using the CNDO/2 approximation, we decided to adopt *Pople's* classical partitioning for the estimation of these energy components:

$$E_{tot} = \sum_A E_A + \sum_{A<B} E_{A-B} + \sum_{A<B} E_{A...B} \cdot$$

The E_{A-B} terms correspond basically to the *chemical bonds* of the molecule and the $E_{A...B}$ terms to the through-space interactions between *non-chemically bonded atoms*.

This arbitrary partitioning of the total energy in mono- and bicentric components has recently been criticized by some theoreticians (*12*), who claimed that such a procedure would lead to "artefacts". Even though these remarks may be mathematically valid, we consider *Pople's* partitioning justified because of the numerous ways in which it can help the chemist to understand and predict the preferred conformations and geometries of a molecular system. Let us discuss some examples.

1. The Case of $F_3P \cdot BH_3$

On plotting the variations of E_{tot}, of $E_b = \sum_A E_A + \sum_{A<B} E_{A-B}$, and of $E_{nb} = \sum_{A<B} E_{A...B}$ as a function of the torsional angle θ around the (P.B) bond in $F_3P \cdot BH_3$, it may be seen (Fig. 3) that ΔE_{nb} and not ΔE_b runs parallel to ΔE_{tot}. This clearly means that the preferred staggered conformation of that complex is due to an

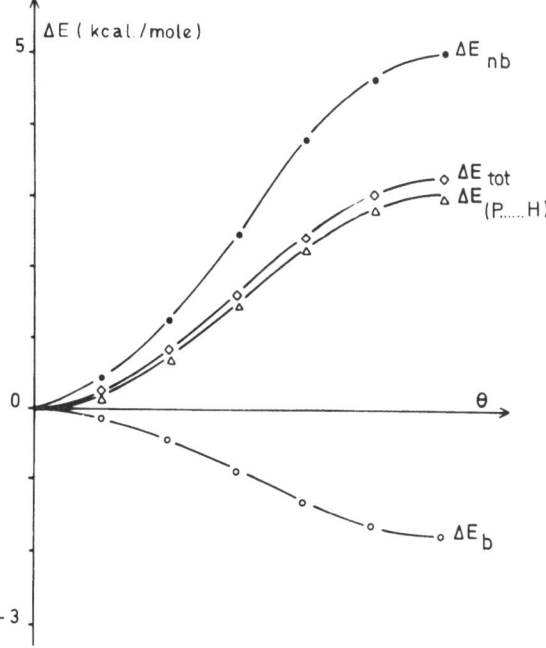

Fig. 3. Relative variations of some components of the total energy of $F_3P \cdot BH_3$ vs. the rotation angle θ around the (P.B) bond

5

optimized balance of "through-space" interactions and not, as would be expected on the basis of purely empirical intuitive ideas, to energy components related to the chemical bonds of the molecular skeleton. More precisely, Fig. 3 shows that the variations of E_{tot} may be reproduced to better than 98% using *one* $\Delta E_{A...B}$ term, i.e. $\Delta E_{P...H}$, which corresponds to a spatial bonding interaction between phosphorus and the three hydrogen atoms of the molecule (Fig. 4).

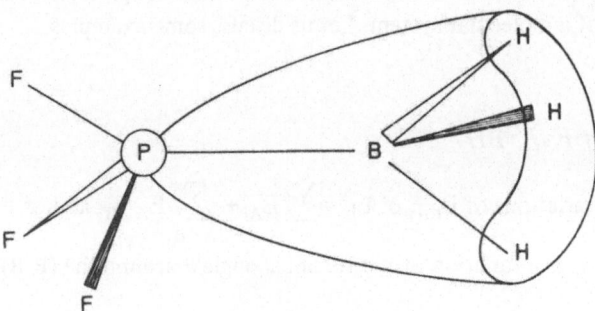

Fig. 4. The through-space bonding interaction term between P and the three H atoms in $F_3P \cdot BH_3$

This example shows that the concept of chemical bond, which is so helpful in the description of the static geometry of a molecule, is quite inadequate to explain the dynamics of intramolecular motions. Moreover (as we shall see below) how can we explain the meaning of the "hyphens" chemists draw between two atoms on the basis of purely topological considerations (i.e. the closer the atoms, the stronger the bond) if calculations reveal large bonding interactions between atoms which seem too far apart to be linked together by a chemical bond? As the next example shows, the concept of chemical bond based on a topological pattern works less well with heavier atoms and becomes quite inefficient in transition-metal complexes. We shall devote a paragraph below to some provocative remarks issued from our works about the "classical" pattern and the related implications of the so-called "chemical bond" in Chemical Education.

2. The Case of Methyl Derivatives

Within the field of Lewis adducts, we naturally studied the conformational analysis of the parent borazane complex, $H_3N \cdot BH_3$, and of its fluorinated derivatives (*13*) and showed that, with respect to their preferred conformation, there is no determinant $\Delta E_{N...H}$ spatial term analogous to the $\Delta E_{P...H}$ term discussed above relative to $F_3P \cdot BH_3$.

This investigation revealed a basic discrepancy between Lewis adducts containing only atoms from the first row of the periodic table and those containing at least one atom from the second row. We therefore initiated a critical examination of the $M(CH_3)_2$ (14) and $M(CH_3)_3$ (15) series in which M could be either a first-row "element" (CH_2, NH, O or CH, N) or a second-row "element" (SiH_2, PH, S or SiH, P).

The main results are as follows: in every case, the computed preferred conformation was identical to that found by the most appropriate experimental technique, i.e. microwave spectroscopy. Moreover, we found it generally true that such conformations are caused by determinant through-space bonding terms only when M is a second-row element, and the $\Delta E_{M...H}$ components in these methyl derivatives were seen to have a determinant role similar in every way to that described for $\Delta E_{P...H}$ in $F_3P \cdot BH_3$. In other words, with respect to preferred conformations, there is interaction between silicon, phosphorus or sulfur and the hydrogens of a methyl group (linked to them by a "normal" covalent bond) and also the hydrogens of a BH_3 group (linked by a coordination bond). This result explains the geometrical modifications (in bond lengths and angles) which occur when the $H_3P \cdot BH_3$ Lewis adduct is methylated step by step on phosphorus (16). The through-space bonding component, $\Delta E_{P...H}$, which determines the preferred conformation of this Lewis adduct, decreases in magnitude upon methylation because the phosphorus atom has to interact with *six* protons in $(CH_3)H_2P \cdot BH_3$ (6), *nine* in $(CH_3)_2HP \cdot BH_3$ and *twelve* in $(CH_3)_3P \cdot BH_3$ (13).

3. The Case of Inorganic Ring Systems

The determinant role of spatial interaction terms has been emphasized in the study of the electronic structure and preferred conformations of some inorganic ring systems.

From 1958, when *Craig et al.* (17) and *Dewar et al.* (18) introduced their intuitive and controversial ideas on the structure of cyclophosphazenes, until recently (19) the extent of the electronic delocalization in inorganic rings has constituted a challenge to both theoreticians and chemists.

An answer to the problem of determining the electronic structure in the ground state of cyclophosphazenes $(NPX_2)_n$ has been supplied by a concerted use of quantum chemistry (19) and the Faraday effect (20), the results of which unambiguously support *Dewar's* island model (18).

As shown in Fig. 5, although the Wiberg indices (21) of the endocyclic (P $\cdot$ N) bonds are large (favoring an important delocalization within the ring), the electron density on the phosphorus atoms is so low that no ring current can exist, hence the delocalization remains "localized" within the PNP units (19).

Moreover, a nonambiguous experimental proof of the validity of Dewar's model was provided by measuring the Faraday effect of a series of $(NPX_2)_{3,4,5}$ compounds (20). For a given X, the molecular magnetic rotations throughout this series are exactly in the ratio $3:4:5$, making allowance for the accuracy of the method. This

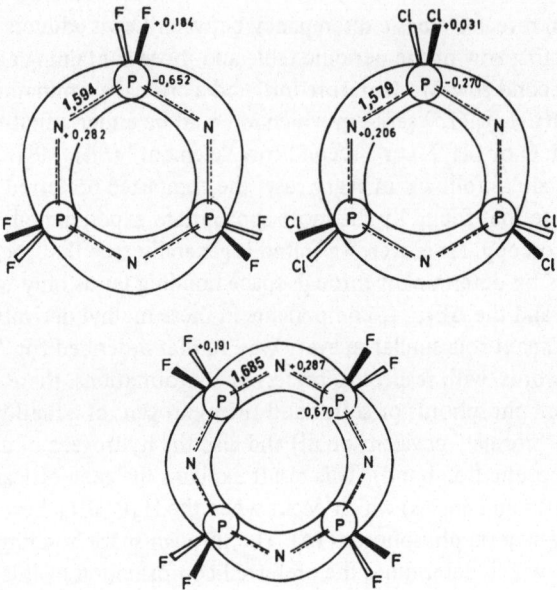

Fig. 5. (P.N.) Wiberg indices and net atomic charges in some standard cyclophosphazenes

result can only be consistent with the existence of individual PNP entities, for if the delocalization in cyclophosphazenes had been of a "benzene-like type", the molecular magnetic rotation would have had to increase exponentially with ring size (i.e. with n) (22).

Now, with the validity of Dewar's model for cyclophosphazenes apparently clearly demonstrated both theoretically and experimentally, a question arose. How is it possible to explain the remarkably high stability of such rings on the basis of localized PNP islands which practically do not interact along the ring skeleton? An answer was provided by quantum chemistry. *Armstrong et al.* (23) and ourselves (24)

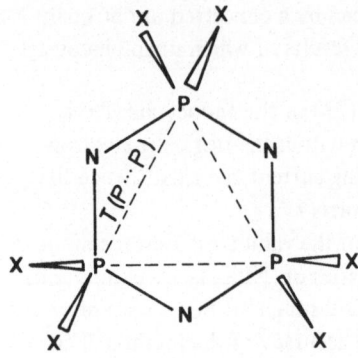

Fig. 6. Transannular bonding (P ... P) terms in (NPX$_2$)$_3$ trimers

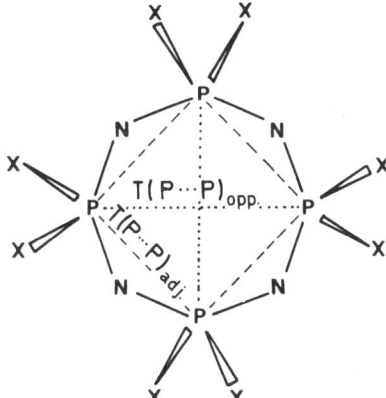

Fig. 7. Transannular bonding $(P \ldots P)_{adj}$ and antibonding $(P \ldots P)_{opp}$ terms in $(NPX_2)_4$ tetramers

have shown that in $(NPX_2)_3$ trimers there is a strong transannular bonding interaction between endocyclic P atoms, each $E_{P \ldots P}$ energy term being about 10% of the $(P \cdot N)$ bond energy (Fig. 6). These through-space interactions appear to be responsible for the interlocking of islands and consequently for the stability of $(NPX_2)_3$ compounds. For $(NPX_2)_4$ tetramers the nature of these interactions is more complicated. The stability and preferred conformations of such rings depend upon the balance of *two* terms, a *bonding* component, $T_{adj.}$, identical to the one mentioned above for the $(NPX_2)_3$ trimer, and an *antibonding* term, T_{opp}. (Fig. 7). The "puckered or not" character of the ring unit through the $N_4P_4F_{8-x}(CH_3)_x$ series was elucidated in that way (*24*).

Such through-space interactions also occur in other inorganic ring systems when two atoms of the second row of the periodic table belong to the ring. We showed, for example, that the "tub" form of N_4S_4 is due to $(S \ldots S)$ bonding terms (*25*) of an order of magnitude relative to the $(S.N)$ bond energy of about 30%, or three times that observed in PN rings. Similarly, it was shown that the preferred "chair" conformation of $(Me_2NAlH_2)_3$ (Fig. 8) is due to strong transannular $(Al \ldots Al)$ bonding interactions amounting to 50% of the $(Al.N)$ bond energy (*26*).

Consequently, it must be emphasized that precautions have to be taken with the conventional rough description of molecules based on the "chemical bond" pattern. In a molecule that contains at least two atoms which do not belong to the first row of the periodic table, the energy and all the monoelectronic properties are literally spread out over the whole molecule. Obviously, the concept of chemical bond, based as it is on the principle of topological proximity, is inadequate on its own for a correct description of the chemical and physical behavior of such a molecule.

Although this paper is devoted to nonmetallic chemistry, we shall go a little beyond the topics we were requested to treat in order to stress the importance of through-space interaction terms by referring to some very recent results from a study now in progress in our group on electronic structures and preferred conformations in the field of transition-metal complexes.

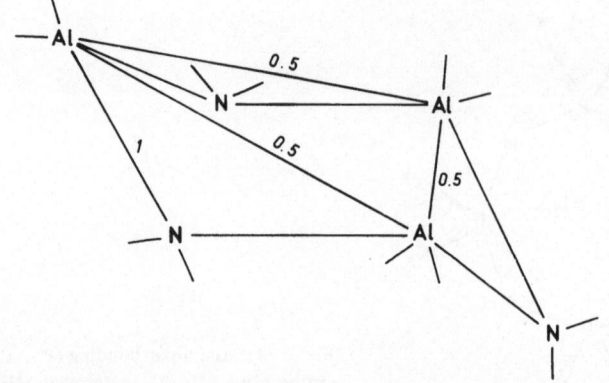

Fig. 8. Transannular (Al ... Al) bonding terms in $(Me_2NAlH_2)_3$

4. Some Preliminary Results in the Field of Transition-Metal Complexes

The synthesis of the trismethylenemethane iron tricarbonyl complex $[(CH_2)_3C]$-$Fe(CO)_3$ was reported by *Emerson et al.* in 1966 (*27*). The geometry of this compound in the gas phase was investigated by *Almenningen et al.* (*28*) using electron diffraction methods. These authors pointed out some structural peculiarities which were not amenable to a simple explanation, in particular, why the hypothetical planar $(CH_2)_3C$ radical is distorted when bound to the $Fe(CO)_3$ conical fragment in such a way that the carbon atoms of the CH_2 groups are displaced *toward* − the iron atom (Fig. 9).

The "attractive and repulsive regions of the bond energy vs. (Fe ... C) distance curve" were invoked (*28*) in order to explain this observation but it was surprising to find a stronger (Fe ... C) bonding interaction for a long distance (2.123 Å) rather than for the shorter distance (1.938 Å). In other words, the question was whether the real electronic structure of the complex was an *"umbrella"* one (in which a fourth Fe−C strong bond would exist along the iron-central carbon direction, this structure being

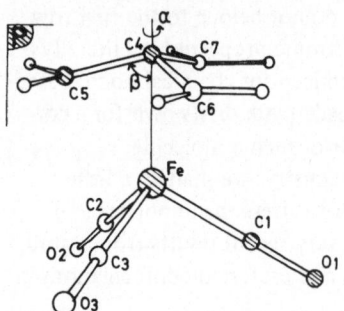

Fig. 9. Geometry and numbering of atoms in $(CH_2)_3C\,Fe(CO)_3$

Table 1. Interatomic distances, bicentric energy components, and Wiberg indices for (Fe, C) pairs in $[(CH_2)_3C]Fe(CO)_3$

	Interatomic distance Å	E_{AB} (u.a.)	W_{AB}
FeC_1	1.810	− 0.31	0.60
FeC_4	1.938	− 0.06	0.21
FeC_5	2.123	− 0.30	0.46

Net atomic charges Q_A
Fe: 0.67; C_1: 0.09; C_4: 0.17; C_5: − 0.30; O_1: − 0.05; H: − 0.01

supported by the classical relationship between interatomic distance and bond strength) or a *"parachute"* one (in which the Fe(O) atom would be at least hexaco-ordinated, i.e. preferentially linked to the carbon atoms of the three CH_2 groups).

A conformational analysis of this complex was performed (*29*) using the extended CNDO method we had proposed for the study of transition-metal complexes (*30*) (*31*) (*32*) (*33*). We were able to show that the electronic structure of $[(CH_2)_3C]Fe(CO)_3$ is really a parachute and not an umbrella, once more using Pople's energy partitioning. The expected simple linear relationship between bond strengths, as measured by E_{AB} or W_{AB} (Table 1), and interatomic distances was shown to be nonexistent owing to the sign of the net atomic charges in the molecule (Table 1). It is noteworthy that such a parachute structure supports the recent work of *Miller et al.* (*34*) concerning the vibrational spectra and force field of the said complex. This example clearly shows that to draw links symbolizing chemical bonds between atoms on the simple basis of interatomic distances may lead to erroneous conclusions both for molecular preferred conformations and even for electronic structures.

Such a peculiarity has also been observed in many other compounds with metal-carbon bonds, especially cobalt-carbon. As part of a study (using the above-mentioned extended CNDO method) of the $AM(CO)_3$ isoelectronic series (AM from η^2-C_2H_4Ni to η^6-C_6H_6Cr) it was shown that, in the η^3-C_3H_5Co derivative, the allyl group was linked to cobalt by means of the two $C(H_2)$ carbons and not predominantly by the C(H) carbon atom, despite the fact that Co−$C(H_2)$ = 2.10 Å as compared with Co−C(H) = 1.98 Å (*35*).

The case of *dinuclear* complexes like $Mn_2(CO)_{10}$ illustrates the important part played by through-space interactions in the stability of such derivatives. Indeed, ex-tended CNDO calculations (*36*), in agreement with earlier SCCC ones (*37*), show that, contrary to what would have been expected according to simple intuitive ideas, the Mn ... Mn interaction, corresponding to the Mn−Mn covalent bond built from the two unpaired 3 d^5 manganese electrons, is equal to − 0.2 a.u. as compared with the *eight* through-space bonding terms that exist between each Mn atom and the four equatorial (relative to the direction Mn ... Mn) CO groups linked to the second Mn. The sum of these interactions is 8 (− 0.08) = − 0.64 a.u. and this contributes more than 75% to the stabilization forces of $Mn_2(CO)_{10}$. In other words, the structure of this complex is due not only to a (Mn–Mn) linkage but also predominantly to (Mn · CO) through-space interactions. Incidentally, such contributions may explain why the eight equatorial CO groups are tilted some $3°4$ *toward* the center of inversion of the molecule, as demonstrated by electron diffraction (*38*).

The facts noted in the four subsections of this chapter suggest that chemists ought to keep an open mind concerning the concept of chemical bonding. Useful as this concept is the field of organic chemistry, it becomes more and more meaningless and inadequate as we move toward inorganic, nonmetallic and, above all, metallic chemistry.

III. The Use of Quantum Calculations Combined with Experiment to Determine Molecular Geometry in the Gas Phase

As stated above, CNDO formalism was able to predict for many methyl derivatives (containing numerous hydrogen atoms) preferred conformations fully identical to those obtained by the most appropriate experimental techniques, electron diffraction and microwave spectroscopy. This was the case, for example, for each term of the $(CH_3)_2M$ (*14*) and $(CH_3)_3M$ (*15*) series. This quantum approach appeared likely to help experimentalists to locate accurately, and in a simpler way than usual, the light atoms – mainly hydrogen – in a molecule.

Indeed, the precise location of hydrogen atoms by means of microwave spectroscopy involves the study of a complete set of isotopic (deuterated) derivatives, which is often costly and time-consuming. An alternative method, based on neutron diffraction measurements, is difficult to carry out for many obvious reasons.

Electron diffraction provides experimental diffraction spectra for comparison with computed spectra obtained from various intuitive geometrical models, but this technique alone is generally insufficient to locate the hydrogen atoms. A quantum approach, on the other hand, indicates the positions of the H atoms, which can then be introduced into the calculation of the theoretical spectra in order to complete the determination of the geometry.

We have combined this kind of work with electron diffraction on many occasions. Let us mention the example of the $Cl_3Al · NH_3$ Lewis adduct we studied with *M.* and *I. Hargittai* (Budapest) and *V. P. Spiridonov* (Moscow) (*39*).

It has sometimes proved possible to determine the location of the hydrogen atoms in a given molecule by calculation when it could not for various intrinsic reasons be found by experiment. This was the case for dimethylsulfoxide $(CH_3)_2SO$ (*40*). Although many experimental studies of its geometry in the gas phase had shown that the three (C–H) bond lengths within each methyl group were significantly different, it had proved impossible to determine precisely why.

Starting with the most accurate geometry of $(CH_3)_2SO$ (which incidentally leads, when compared to the others, to the best value of the total energy), we drew the three-dimensional potential surface $E = f(\alpha, \beta)$ describing the variations of the total energy as a function of the rotations α and β around the two (C–S) bonds. Fig. 10 shows that the potential well is not observed for the (60, 60) conformation (*41*), although it would have been expected if $(CH_3)_2SO$ had been of C_{2v} symmetry. The two hydrogens, which would have lain within the CSC plane in a (60, 60) form, are actually tilted 3°54 out of that plane and toward the oxygen, which would explain the experimental set of three different (C–H) bond lengths. In this way we found a geometrical feature not attainable by any experimental study, even neutron diffraction. Moreover, the way in which this tilt occurs shows that a $(S \cdot O)$ dative bond appears to be less bulky than the lone pair of a sulfur atom in a sulfoxide, and this insight may be helpful for many purposes.

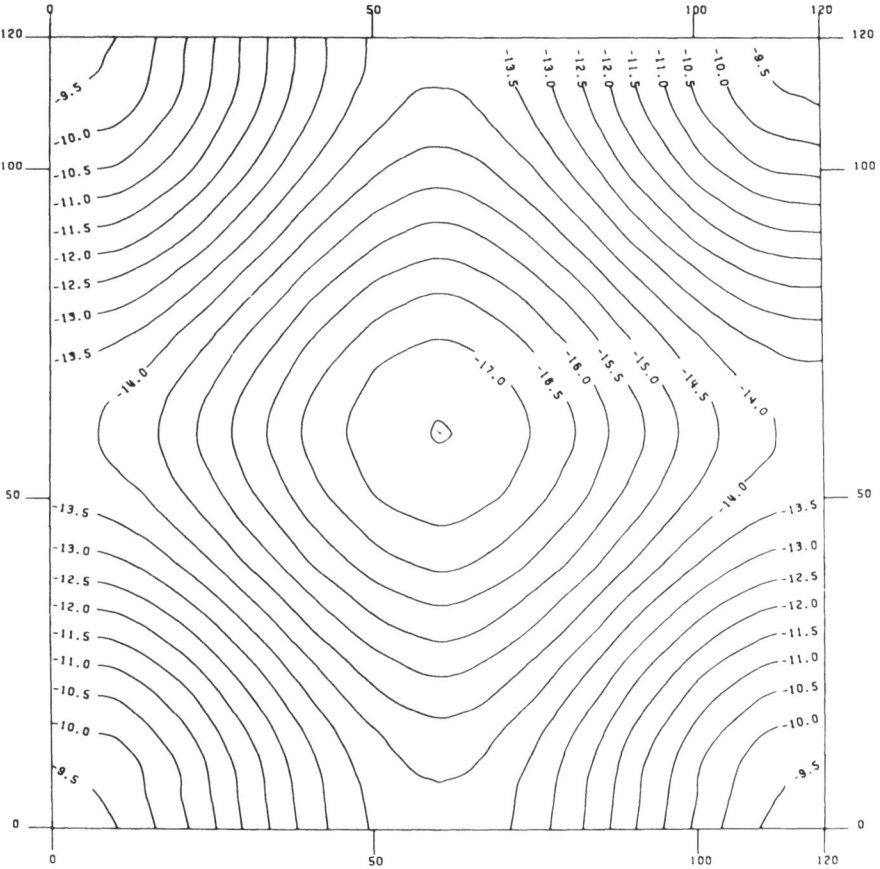

Fig. 10. Map of isoenergy curves for $(CH_3)_2SO$

13

Further, some CNDO calculations on the protonation path of the most basic cyclophosphazenes, namely $N_3P_3(NMe_2)_6$ (42), have shown that such a protonation deeply affects the endo- and exocyclic (P.N) bond lengths, inducing inversion of their relative stabilities, so leading to the conclusion that the PN ring of a protonated species has a nonregenerative character. It was therefore predicted that, in view of the increase in the endocyclic (P·N) bond lengths and the decrease in the corresponding exocyclic ones upon protonation, the cyclophosphazenic ring of the protonated molecule could be broken by electrophilic attack. All these results (variations of (P·N) bond lengths upon protonation and reactivity of protonated species) have been supported by recent X-ray and chemical work (43–45).

A fourth example of the value of combining theory with experiment in order to carry out the conformational analysis of a molecule is the case of the complex $(CH_3)_2SO·BF_3$. An X-ray crystallographic study had shown (46) that the preferred conformation of this complex in the solid state would be that visualized in Figure 11. A study of the vibrational spectra and force field of this molecule in dilute solution showed that this preferred conformation could not be transposed to the liquid state because there exist in solution some (F ... H) hydrogen bonding interactions which, not being affected by dilution, must consequently have an intramolecular character. Such (F ... H) interactions cannot be understood from the solid-state preferred conformation so that we had to reexamine the conformational analysis of $(CH_3)_2SO·BF_3$.

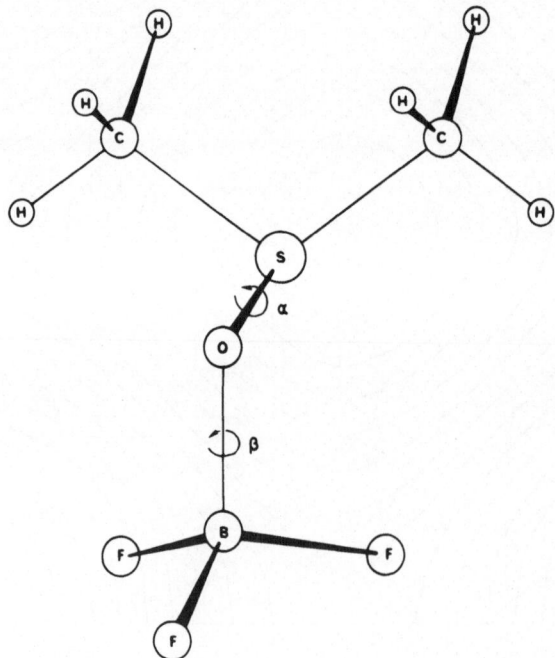

Fig. 11. McGandy's conformation in the solid state for $(CH_3)_2SO·BF_3$

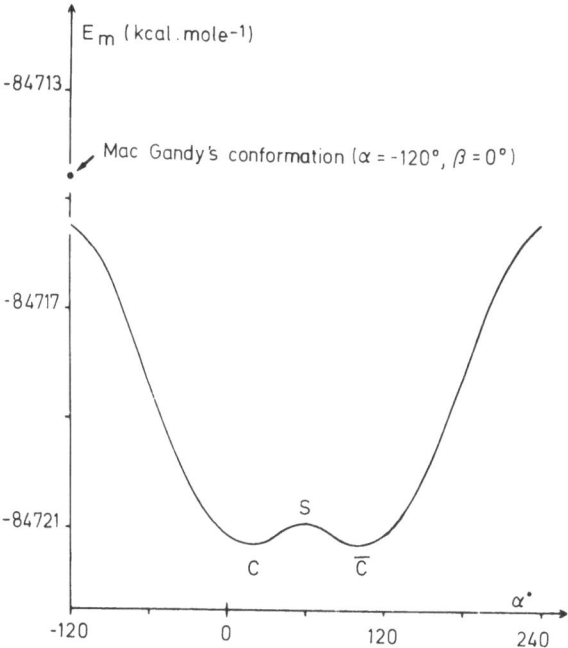

Fig. 12. Variation of the total energy of $(CH_3)_2SO \cdot BF_3$ as a function of the rotational angle α

Fig. 12 shows that, within the CNDO approximation, the X-ray conformation is no longer adopted in solution. In fact, the complex oscillates around the S conformation (Fig. 13) from the gauche C to the symmetrical $\bar{C}$ conformation (Fig. 14). A close look at this motion indicates that it may be decomposed into a simultaneous

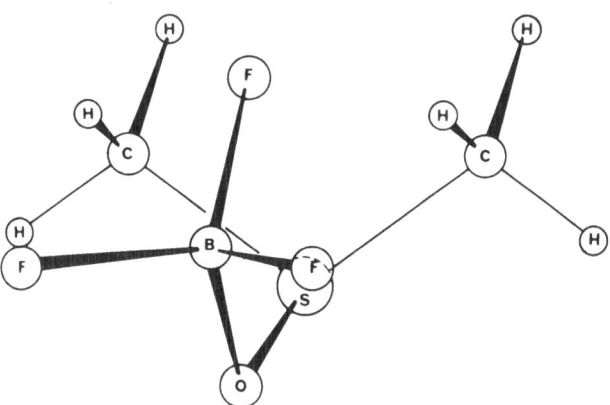

Fig. 13. The S conformation of $(CH_3)_2SO \cdot BF_3$

15

rocking around the $(S \cdot O)$ bond and *rolling* around the $(O \cdot B)$ bond. This *"rock-and-roll"* internal motion, predicted by pure quantum calculations (*47*), allowed us to interpret the i.r. results mentioned above.

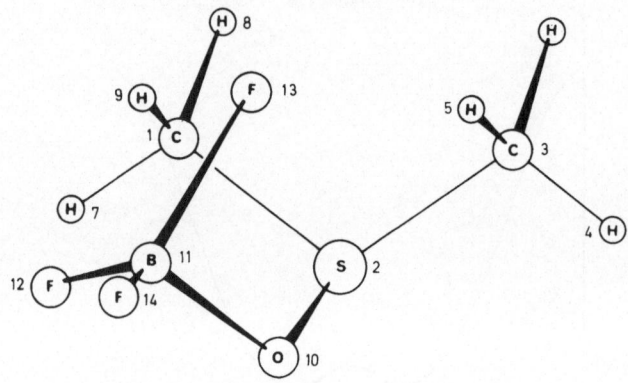

Fig. 14. The C conformation of $(CH_3)_2SO \cdot BF_3$

IV. The Assumption of a Standard Geometrical Model: Its Use when Quantum Chemistry Is Applied to Inorganic Chemistry

We have presented above some examples in which the theoretical conformational analysis of a molecule had to be supported by — or could help to support — an accurate knowledge of the static geometry in the gas phase.

A fundamental problem for the chemist is to elucidate the factors that determine the *trends* of physical and chemical behavior within *large series* of "non-academic" (i.e. complicated) molecules. A knowledge of the precise geometry of *each* term of the series might seem necessary before one can approach the analysis of such trend by means of quantum chemistry. If this were so, the help provided by calculations would soon be limited by the fact that the geometry of many molecules in the gas phase cannot be determined by experiment. For this reason we decided in 1972 to test within the field of inorganic chemistry the validity of the *standard geometrical model* so successfully applied in organic chemistry by *Gordon* and *Pople* (*11*). This pattern is based on the assumption that bond angles and lengths remain unaltered within a series of homologous compounds, variations in behavior being revealed by the SCF procedure.

We had been working with Lewis adducts for many years, so that the prediction of the basic properties of bases with respect to a given Lewis acid was extremely important to us. We therefore decided to start with a general quantum approach to basicity in the gas phase within the assumption of a standard geometrical model. Our first task was to reproduce the basicity scales in the gas phase toward H^+, as provided by such techniques as ion-cyclotron-resonance spectrometry or photoelectron spectroscopy, and to distinguish the main factors such scales depend on. We successively examined the amine (48), alcohol (49), phosphine (50) (51), and thiol (52) families and obtained the following results.

1. First of all, quantum calculations allow one to predict basicity scales in agreement with experiment *provided that the calculations are performed on the preferred conformation of the isolated molecule.* If this is not done, a given term within a consistent series may jump from one rank to another as a function of the conformation used for the calculations. The determinant role of preferred conformation on any property (barrier to internal rotation and inversion, dipole moment, first adiabatic ionization potential, acidity and basicity in the gas phase, energy of complexation to BF_3, etc.) was clearly demonstrated. We further show the importance of the role of preferred conformation in explaining some of the anomalies in *Drago's* systematics.

2. Once the preferred conformation of each term of the series has been settled, the basicity (and acidity) scales provided by CNDO calculations on the assumption of standard geometrical models are *identical* to those obtained from i.c.r. and u.p.s.. Quantum calculations even allowed us to predict the basicity of some Lewis bases which could not be experimentally determined because of side-reactions occurring during the measurements.

3. In addition it is possible to explain the inversion about the acidity scale of aliphatic alcohols observed on passing from the gas phase to aqueous solutions. *Jano's* formalism (53) for solvation energy was successfully used on this occasion.

V. A Conformational Explanation of Some Discrepancies in Drago's Systematics

The double-scale four-parameter enthalpic equation proposed in 1965 (54) and successfully developed by *Drago et al.* seems to be the best tool so far available for correlating and predicting the formation enthalpies of Lewis adducts in the gas phase or, if really necessary, in solution.

However, *Drago's* work contains some cases where a discrepancy is observed between the calculated and experimental ΔH values. In particular, although the adducts built from methylated Lewis bases (like NMe_3 or PMe_3) are amenable to *Drago's* pattern, their isologous ethyl derivatives are not, and no clear explanation has been proposed for this observation.

Fig. 15. The preferred (60, 60, 60) conformation for isolated NEt_3 and PEt_3

In the light of our earlier work on amines and related bases, which had revealed the determinant role of preferred conformations on basicity scales, we guessed that such conformational factors could be responsible for the above-mentioned discrepancy. We therefore examined from a conformational point of view the modifications that can be induced in NEt_3 or PEt_3 by the approach of a standard Lewis acid, taken as $AlMe_3$. Starting with the two "blocks" (acid and base) far enough apart to be considered noninteracting [the conformations of NEt_3 or PEt_3 and of $AlMe_3$ being their preferred ones when isolated (Fig. 15)], we let these two blocks approach along a common C_3 axis, determining at each step a staggered conformation for the hypersystem. Then we computed *the new preferred conformation of the base* for each $d(N ... Al)$ and $d(P ... Al)$ within the hypersystem, assuming that no modification

Table 2. Energy of the preferred conformation of the Et_3N block in $Et_3N \cdot AlMe_3$ as a function of the distance (N ... Al).

Conformation →	(0, 0, 0)	(30, 30, 30)	(60, 60, 60)	(120, 120, 120)
$d(N ... Al)$				
↓				
1.5	− 59980.63	− 60224.58	− 60070.17	− 60333.25
2.0	− 60562.88	− 60627.43	− 60591.43	− 60585.58
2.5	− 60634.38	− 60636.42	− 60602.40	− 60551.49
3.0	− 60526.97	− 60573.76	− 60542.17	− 60489.48
3.5	− 60579.72	− 60525.17	− 60504.08	−
4.0	− 60495.92	− 60499.26	− 60488.13	−
4.5	− 60482.00	− 60482.99	− 60487.65	−
5.0	− 60475.76	− 60480.82	− 60483.98	−

Table 3. Energy of the preferred conformation of the Et_3P block in $Et_3P \cdot AlMe_3$ as a function of the distance (P ... Al).

Conformation →	(0, 0, 0)	(30, 30, 30)	(60, 60, 60)	(120, 120, 120)
d(P ... Al) ↓				
1.5	− 57276.08	− 57363.90	− 57314.99	− 57415.61
2.0	− 57676.92	− 57700.77	− 57687.13	− 57688.92
2.5	− 57735.50	− 57737.63	− 57725.28	− 57705.48
3.0	− 57676.75	− 57677.16	− 57667.88	− 57646.14
3.5	− 57607.55	− 57610.26	− 57605.85	−
4.0	− 57564.63	− 57569.28	− 57568.73	−
4.5	− 57545.72	− 57551.39	− 57553.06	−
5.0	− 57539.13	− 57545.21	− 57547.88	−

occured on $AlMe_3$; this "new" conformation at each step was distinguished by means of the minimal energy criterion (55).

As a measure of simplification, the only total energy values necessary for the discussion are assembled in Tables 2 and 3. The meaning of the symbol (α, α, α) used to describe the preferred conformation of NEt_3 or PEt_3 blocks is as follows: the (0, 0, 0) standard is visualized (Fig. 16) and consequently the (α, α, α) notation is related to a form in which the three ethyl groups are simultaneously rotating anti-clockwise by α (the C_3 symmetry being maintained) around the corresponding (N–C) or (P–C) bonds. For example, the "propeller" preferred conformation of free NEt_3 or PEt_3 (Fig. 15) will be labelled (60, 60, 60).

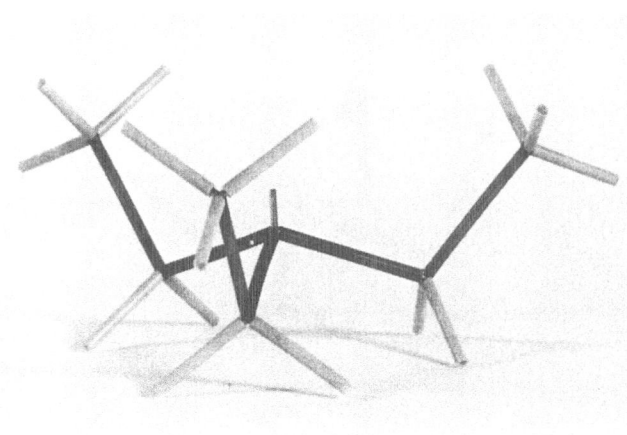

Fig. 16. The standard (0, 0, 0) conformation

Thus it appears from Tables 2 and 3 that (1) NEt_3 and PEt_3 remain in a (60, 60, 60) form as long as $d(N ... Al)$ and $d(P ... Al)$ are larger than or equal to 4.5 Å; (2) on the other hand, as soon as $d(N ... Al)$ and $d(P ... Al)$ reach 4.0 Å, the (60, 60, 60) form vanishes, indicating that the two blocks begin to interact, even at such long distances; the (30, 30, 30) form (Fig. 17) appears to be preferred in both cases; (3) on further decreasing $d(N ... Al)$ and $d(P ... Al)$, the (30, 30, 30) conformation remains the preferred one for $Et_3P \cdot AlMe_3$ to 2.0 Å (Table 3), whereas the (0, 0, 0) and then the (30, 30, 30) form are again observed under the same conditions for $Et_3N \cdot AlMe_3$ (Table 2). In other words, for the two Lewis adducts where $d(N \cdot Al)$ and $d(P \cdot Al)$ were measured by microwave spectroscopy and found to be equal to 2.1 and 2.43 Å, respectively (56), the preferred *real* conformation of NEt_3 and PEt_3 is no longer (60, 60, 60) but (30, 30, 30).

We felt that this result could explain *Drago*'s discrepancies (Et/Me). We know that basicity is critically dependent on conformation. Consequently, the *real* basicity of NEt_3 or PEt_3 versus $AlMe_3$ will differ from the basicity predicted from the two isolated bases. Such a conformational explanation of *Drago*'s difficulties can be generalized to many other cases where conformational modifications have more obviously to be taken into account. For example, if the calculated and experimental ΔH values for a complex like the 1,4-dioxan $\cdot ZnMe_2$ adduct are not in agreement (57), this must be due to the fact that free 1,4-dioxan is in "chair" form when it ought to be in "boat" form to give a 1 : 1 adduct with $ZnMe_2$ (58).

In concluding this section, we wish to emphasize that extreme care is necessary when predicting the basicity of donor molecules from experimental or theoretical studies performed on these molecules in their free state. Such an approach appears to be correct when the acceptor moiety is H^+ but may lead to error as the Lewis adduct becomes bulkier, since its "volume" induces dramatic conformational changes with

Fig. 17. The (30, 30, 30) conformation

consequent huge modifications of the *effective* basicity in the antagonist bases. This may explain the failure to define a general and unique basicity scale that would be valid for any the Lewis acid.

VI. A Conformational Approach to the Reactivity of Lewis Bases Containing Competitive Nucleophilic Sites

Considerable use has been made of the power of borane Lewis acid to determine the basicity of group V donors (*59*). Some studies have demonstrated that phosphorus in PR_3 is a stronger base than nitrogen in NR_3 when BH_3 is used as the reference acceptor molecule (*60–63*). This problem has been tackled by studying the behavior of the two potential donors when both are set up in the same molecule. Typical examples occur in the work of *Reetz* (*64*), who cites evidence for the donor capacity of the phosphorus atom in $(R_2N)_3P$, in various papers by *Holmes* (*65*) and *Parry* (*66*) dealing with different classes of phosphines such as $(R_2N)_xPF_{3-x}$ and their borane adducts, and in the recent work of *Jugie et al.* (*67*) centered on the study of halogen-oborane adducts of aminophosphines.

In all the above-mentioned examples, the greater capacity of phosphorus is explicable in terms of the "double-bond" character of the $(P \cdot N)$ bond. In order to exclude such a possibility of π back-bonding, *Miller et al.* (*68*) used donor molecules in which the P and N atoms were separated by a CH_2 group. Their n.m.r. results suggest that, even so, the coordination of BH_3 occurs on the phosphorus site and not on the nitrogen site.

Surprisingly, the reverse situation is found in the case of the aminophosphine $L = (Et_2NCH_2)_3P$ (*69*), where it was shown by n.m.r. (*69*) and calorimetry (*70*) that BH_3 preferentially attacks the three nitrogen atoms in the first step, readily giving the L.3 BH_3 adduct, the fourth BH_3 being grafted on P, giving L.4 BH_3, in quite a poor yield ($\cong 60\%$) and this only under drastic conditions (*69*). In order to interpret such a tricky observation, *Jugie et al.* invoked steric hindrance, but without any definite proof of this assumption. *Jugie* therefore urged us to look for a quantitative explanation for this anomaly by carrying out a theoretical conformational analysis within the field of the CNDO approximation.

In the first stage we dealt with the determination of the preferred conformation for the free L (*71*). As the geometry of this aminophosphine in the gas phase is unknown, we adopted the following model: CPC angles and (P–C) bond lengths, respectively, were taken as $100°$ and 1.846 Å, as in PMe_3 (*72*), while $d(C–N) = 1.472$ Å was taken from the microwave geometry of NMe_3 (*73*). *Gordon* and *Pople's* standard geometrical model was used for the rest of the structure. Assuming that the dimethylamino groups are fixed in the classical (60, 60) conformation (*14*), the conformational analysis of L can be limited to the drawing of the $E = f(\theta, \phi)$ hyper-

21

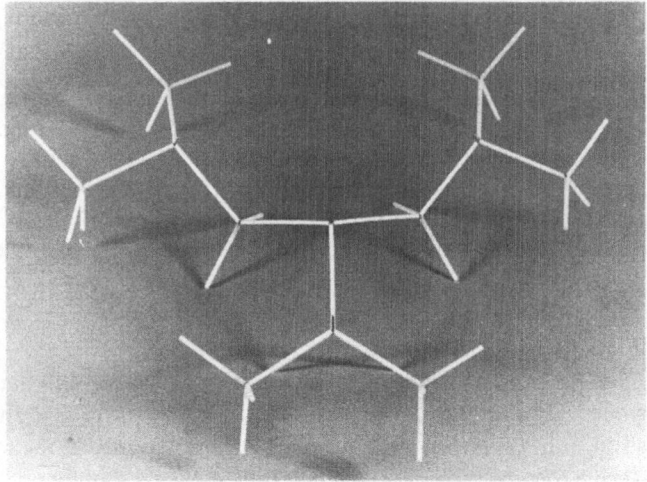

Fig. 18. The standard (0, 0) conformation for the free aminophosphine L

surface, with θ and ϕ the anticlockwise rotations (for an observer located on P and C) around the (P–C) and (N–C) bonds, respectively. In these conditions, the standard (0, 0) conformation is as in Fig. 18.

Fig. 19 gives the map of the isoenergy curves for L. This Lewis base can exist in two preferred conformations, L_1 and L_2:

$$L_1: \theta = 40°; \quad \phi = 112°; \quad E = -78334.4 \text{ kcal. mole}^{-1}$$
$$L_2: \theta = 318°; \quad \phi = 127°; \quad E = -78333.8 \text{ kcal. mole}^{-1}.$$

It may be noticed that these two conformations are frozen, as the potential barrier to pass from L_1 to L_2 or from L_1 to L_2' is unfavorable, being 17 kcal. mole^{-1}, and more than 40 kcal. mole^{-1}, respectively.

The spatial arrangement of the atoms in L_1 and L_2 (Figs. 20 and 21) suggests that *the location of the methyl groups encages the phosphorus lone pair and consequently makes the three nitrogen lone pairs directly accessible to BH$_3$.*

Theory thus seems able to explain the origin of the experimental observation concerning the preferred fixation of BH$_3$ on N with respect to P. But we had to go further and try to understand why the graft of the fourth BH$_3$ on L.3 BH$_3$ is so difficult. Therefore, we carried out the theoretical conformational analysis of the L.3 BH$_3$ complex. Starting with a standard (0, 0) conformation in which the three BH$_3$ are linked to the N atoms of L in a staggered conformation with respect to the bonds issued from each N[d(B–H) = 1.211 Å; $\widehat{\text{NBH}}$ = 105° · 32] at a distance (B·N) of 1.638 Å, as in Me$_3$N · BH$_3$ (74), we drew the E = f(θ, ϕ) hypersurface as before (Fig. 22). It becomes apparent that the L.3 BH$_3$ complex exists in only *one* (38, 355) preferred conformation (E = -90, 261.2 kcal. mole^{-1}) (Fig. 23).

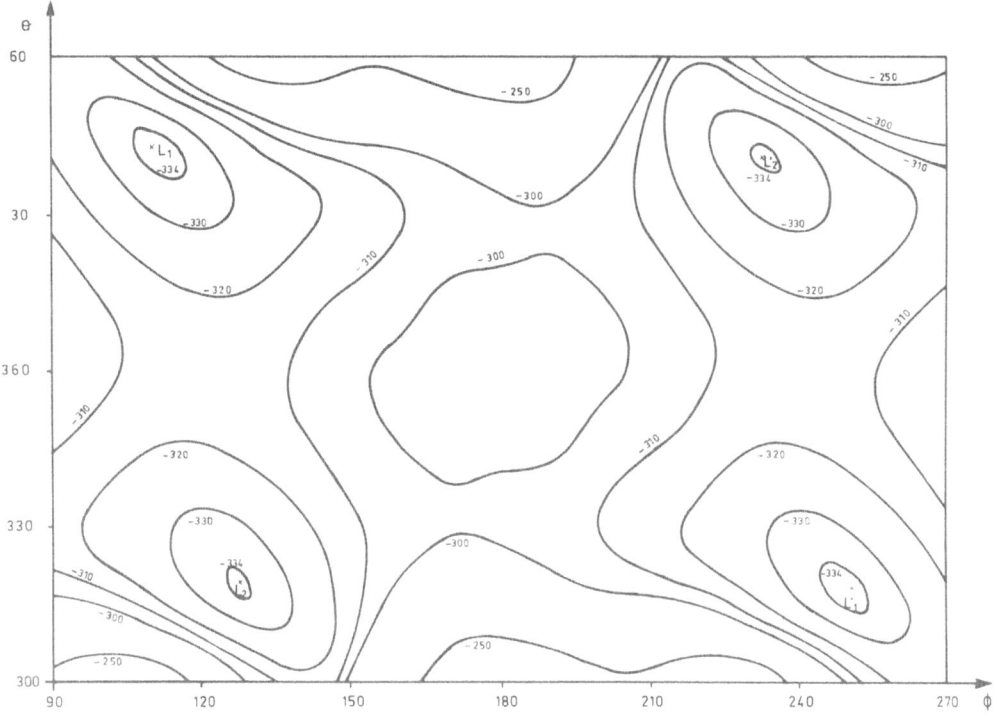

Fig. 19. Map of isoenergy curves (of $-78,000$ kcal$\cdot$mole^{-1}) for the free aminophosphine L

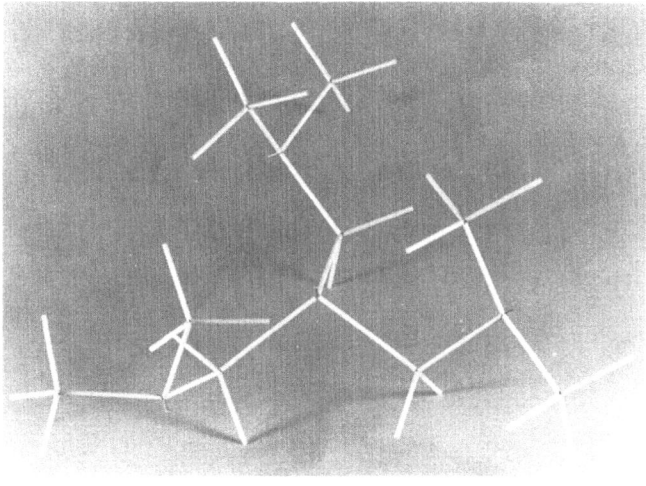

Fig. 20. The preferred conformation L_1 of L

23

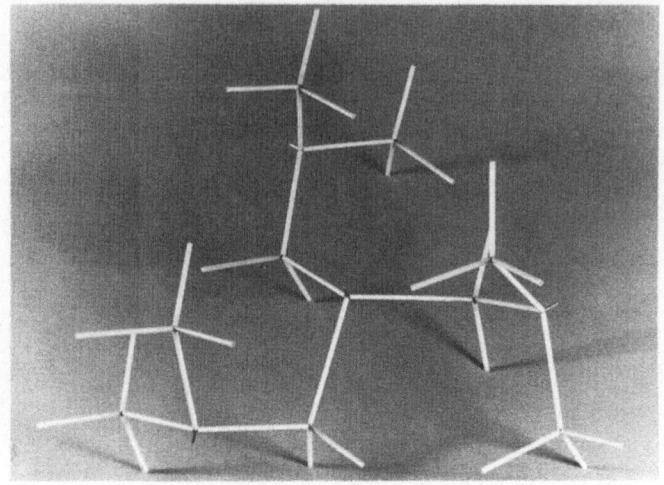

Fig. 21. The preferred conformation L_2 of L

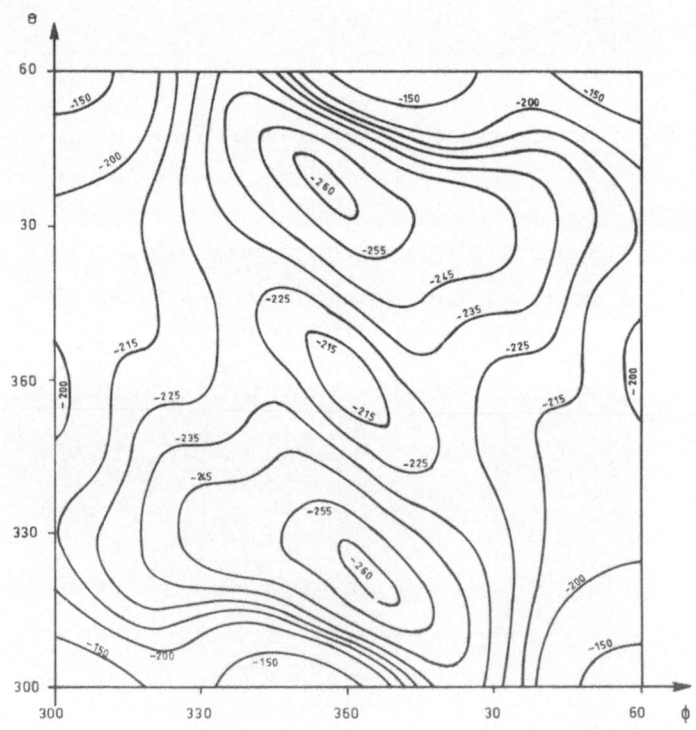

Fig. 22. Map of isoenergy curves map (of $-90,000$ kcal $\cdot$ mole^{-1}) for the complex $L \cdot 3$ BH$_3$

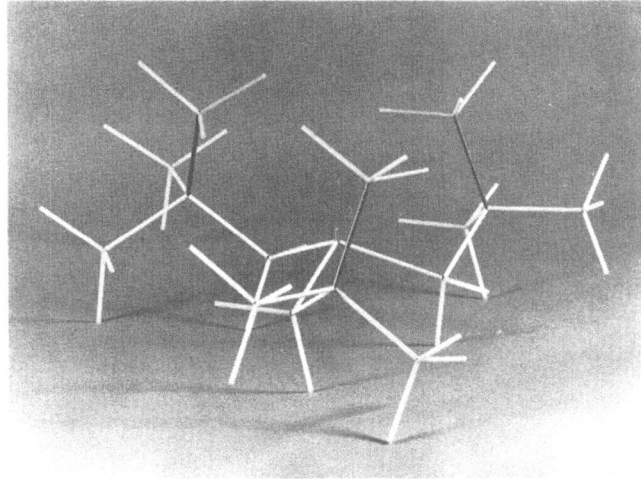

Fig. 23. The preferred conformation of the $L \cdot 3\,BH_3$ complex

Thus, the $(B \cdot N)$ bonds are approximately directed toward the lone pair of phosphorus, in which case, the three BH_3 groups play the same part as the three CH_3 groups in L. This explains why the graft of BH_3 on P is so difficult.

We would stress that, beyond this theoretical understanding of a given experimental observation, the quantum approach shows that the reaction path under coordination of three BH_3 to L cannot be described in any way through smooth intramolecular rotations. A study of Figures 19 and 22 indicates, indeed, that the passage from L_1 (or L_2) to the $(38, 355)$ preferred conformation of $L.3\,BH_3$ is energetically forbidden, the barrier for that rotation being much too high. Our theoretical analysis suggests that this passage occurs through an activation state which must involve an inversion of the phosphorus environment (*71*). This assumption has of course to be supported by further experiment.

VII. A CNDO Approach to Barriers to Internal Rotation in Some Transition-Metal Complexes

We have illustrated the efficiency of the original CNDO/2 formalism for the conformational analysis of nonmetallic compounds and briefly indicated in Section III that an extension of this quantum technique, designed to be applicable to transition-metal complexes (*30–33*), allowed us to generalize the importance of through-space bonding interaction terms in stabilizing dinuclear complexes like $Mn_2(CO)_{10}$ (*75*). We now describe some results obtained with the extended CNDO formalism concerning preferred conformations and related parameters within the field of mononuclear coordination derivatives.

The first example concerns the "free rotation or not" of PF_3 groups in some $M(PF_3)_n$ complexes (*32*). In $Ni(PF_3)_4$, *Almenningen* (*76*) fitted the experimental electron diffraction data by assuming that the four PF_3 groups were freely rotating, while *Marriott* (*77*) found (also by electron diffraction) a preferred conformation with a torsional angle $\tau = 40°$ ($\tau = 0°$ when the P–F bonds of one PF_3 are in an eclipsed conformation with respect to the Ni·P bonds linking nickel to the other three PF_3 groups). Taking $d(Ni·P) = 2.099$ Å (*76*), our calculations (*32*) lead to a minimum of total energy for $\tau = 60°$. We should, however, stress that the difference in energy between the eclipsed and staggered forms is 6.8 kcal. $mole^{-1}$, i.e. four times the value (1.7 kcal. $mole^{-1}$) calculated when only one PF_3 group is rotating. This shows that there is no interaction between the rotations of the different PF_3 groups. The rather low value for such a rotation barrier indicates that the four PF_3 in $Ni(PF_3)_4$ are virtually freely rotating. Moreover, a statistical analysis of the number of PF_3 groups, characterized by a torsional angle τ as a function of the energy of the corresponding conformation, leads to a "most probable" model in which τ equals precisely 40°. As electron diffraction is known to determine only *average* interatomic distances, the last model is in perfect agreement with the "conformation" proposed by *Marriott*, but it is not surprising that *Almenningen* had found the PF_3 groups to be freely rotating. On the basis of extended CNDO calculations we were thus able to reconcile the two experimental results provided by electron diffraction.

As regards $Fe(PF_3)_5$, our calculations (*32*) indicate that the equatorial PF_3 groups are freely rotating (rotational barrier $\simeq 0.3$ kcal. $mole^{-1}$) while the axial PF_3 groups are staggered with respect to the equatorial (Fe·P) bonds ($\tau = 60°$), the eclipsed/staggered energy gap being 8 kcal. $mole^{-1}$. It may be noted that only one ^{31}P n.m.r. signal is observed for this compound, but the experiment was carried out at room temperature and no VT process was performed (*78*).

For $Cr(PF_3)_6$, *all* the PF_3 groups were found to be in free rotation (rotational barrier $\simeq 0.3$ kcal. $mole^{-1}$).

The second example is related to the rotational barrier around the C_3 axis of benzenechrometricarbonyl, $C_6H_6Cr(CO)_3$. Our calculations (computing time on an IBM 370/168 $\simeq 10$ min) had shown that this barrier would be less than 0.4 kcal. $mole^{-1}$ (*79*). The conformational analysis of this molecule was concurrently performed by *Schäfer* (*80*), using electron diffraction. It appeared that, contrary to chemical intuition but in excellent agreement with our own calculations, the phenyl group could be considered as freely rotating around the C_3 axis.

This example clearly shows that the use of a quantum technique, which has been shown to be reliable and suitable *within a given area of chemistry,* can be really powerful while economizing in time and money.

VIII. The Use of CNDO Methods to Optimize Molecular Geometries

Work carried out by *Pople et al.* and by ourselves in the fields of organic and inorganic chemistry has helped to show that bond lengths and bond angles can be optimized on the basis of the minimal-energy criterion, provided that the molecules studied are *"non-conjugated"*, i.e. are *electronically localized* systems. We have already shown in Section II that a satisfactory agreement exists between computed and experimental $d(P \cdot B)$ values for $F_3P \cdot BH_3$. We now wish to give further examples of the use of quantum calculations to dermine geometrical parameters, emphasizing the cases where experiment seems, for reasons we develop below, unable to provide these parameters.

1. The Case of Tertiobutylphosphines

The tertiobutyl derivatives of P(III) exhibit quite exceptional chemical and physical behavior amongst the phosphine family (*81*). This behavior has offered a challenge to both the theoretician and the chemist since the seventies, when these molecules were synthesized and studied, mainly by *Schmutzler et al.* Their peculiar behavior was generally explained by postulating a considerable increase in the valency angles on phosphorus (due to the bulky character of tertiobutyl groups) with respect to the $100°$ value which characterizes the whole set of nonfluorinated P(III) compounds. The only experimental indication available in 1971 had been obtained by n.m.r., and *Stelzer* and *Schmutzler* (*81*) proposed CPC angles equal to $109°$ in the tritertiobutyl-phosphine (TTBP). Work was then started with the aim of providing direct evidence for the geometrical structure of such molecules in the gas phase in order to support the indirect n.m.r. results on the CPC value.

From 1972 to the present, samples of TTBP and related derivatives have been sent by *Schmutzler* and ourselves to many experts in electron diffraction or microwave spectroscopy but, despite this, the molecular geometry of TTBP still remains unknown. From the long discussions we had with these experts, it appears that the main reasons for this failure are as follows: the TTBP molecule contains 27 hydrogen atoms and it would have been tedious to prepare the complete set of deuterated species and analyse them by means of microwave spectroscopy, which would have been essential to obtain an unambiguous geometry. As for electron diffraction, the main difficulty arose from the fact that no simple intuitive model could be built to fit the experimental spectrum. We shall see why later.

We therefore tried to perform a complete optimization of the bond lengths and angles of TTBP by means of a pure theoretical approach (*82*). These results are presented in Table 4. It may be noticed that the optimized CPC angle is equal to $108°2$, in good agreement with n.m.r. predictions. Moreover, the three tertiobutyl groups [which are in the *Calder* form (*82*) and not in the LEM form (*15*) Fig. 24] are in-

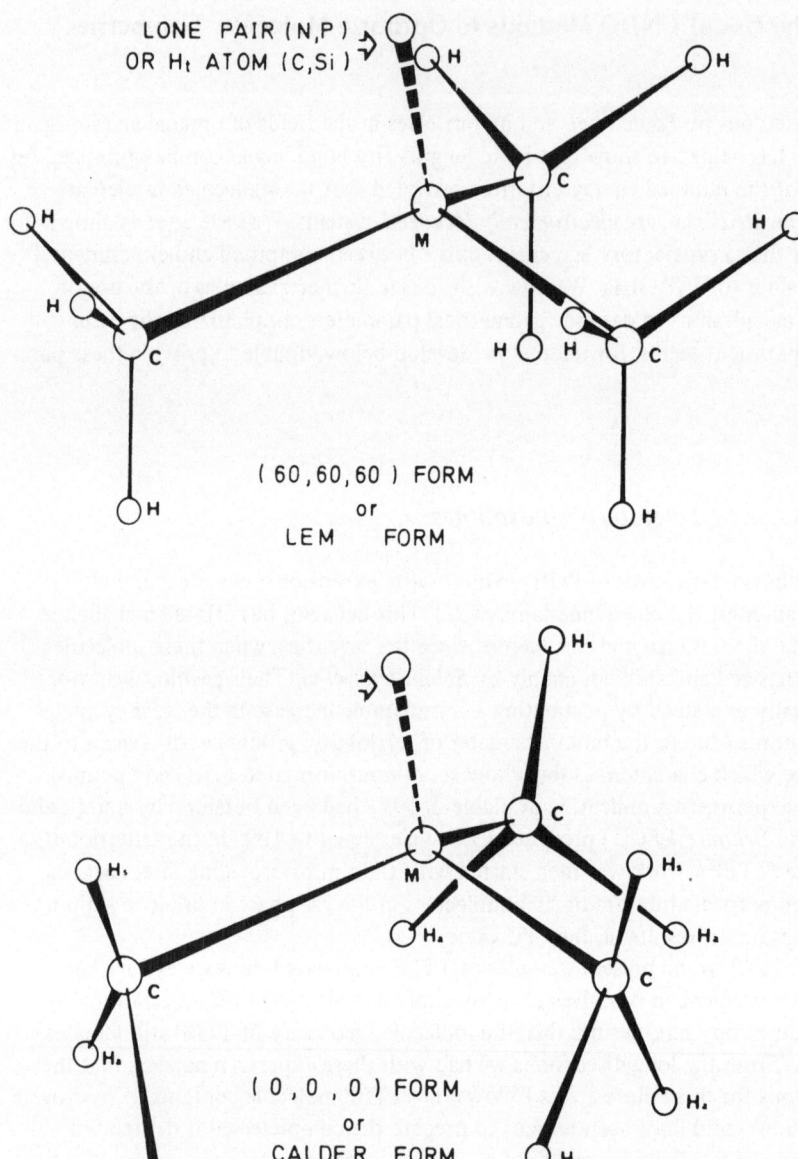

Fig. 24. LEM and CALDER conformations for $(CH_3)_3M$ molecules

Table 4. The optimized geometry of P(t-Bu)$_3$

d(C-C) Å	1.548 (fixed)
d(C-H) Å	1.111 (fixed)
d(P-C) Å	1.846
∢ CCC°	112.1
∢ CCH°	113.1
∢ CPC°	108.2
gear angle°	36

volved in a molecular gearing defined by a 36° twist angle around the (P–C) bonds (Fig. 25). On scrutinizing Figure 25, it is easy to understand why it was impossible to predict intuitively a model to fit the electron diffraction spectra.

On passing from *tri*tertiobutylphosphine to related *di*tertiobutyl compounds like P(t-Bu)$_2$X (X = H or F), a decrease in the CPC angles from 109° to 100° would be expected owing to the smaller number of bulky groups linked to phosphorus. In order to verify this statement, we started a full optimization of the two molecular geometries in question (*83*). Surprisingly, our CPC values were around 114°, contradicting the above assumption. No experimental investigation of these geometries had been performed, either by electron diffraction or by microwave spectroscopy, so we had to search for the structure of a complex in which a P(t-Bu)$_2$X phosphine would be involved as a ligand in such a way that it could be considered only slightly perturbed – or even not at all – upon coordination. This is the case for the *trans*-NiBr$_2$-[P(t-Bu)$_2$F]$_2$ square planar complex, in which the two phosphine ligands are distant

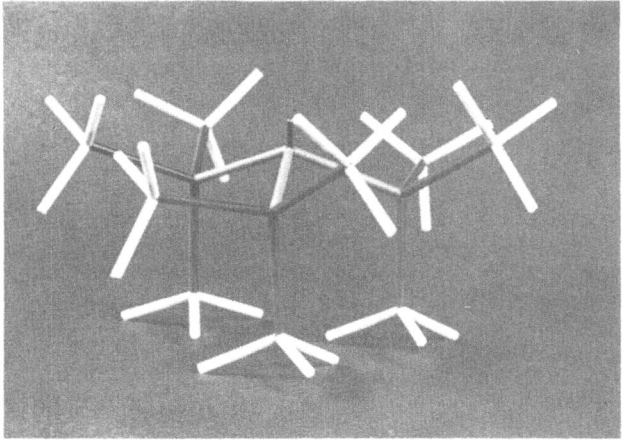

Fig. 25. The preferred conformation of P(t-Bu)$_3$

enough to be in the required situation. This complex was studied by *Sheldrick* and *Stelzer* (*84*) by means of X-ray crystallography, and the CPC angle was found to be (113° 8 ± 0° 6), which agrees well with the calculated angle.

Thus, contrary to chemical intuition, the CPC angle does not decrease regularly with the number of tertiobutyl groups and its maximum value is found for the *ditertiobutyl* derivatives. This curious and unexpected result allowed *Schmutzler et al.* to reinterpret the apparently incomprehensible physicochemical data they had previously collected on the tertiobutylphosphines.

2. The Case of $H_2N\text{-}PH_2$

The molecular geometry of this standard aminophosphine had never been experimentally investigated. Thus it was tempting to perform an optimization of its geometry by means of as nonempirical a method as possible in order to give quantitative support to experimental evidence (i.r., n.m.r., and reactivity) which seemed to suggest a *pyramidal* and not a *planar* arrangement around the nitrogen atom in many aminophosphines (*85–89*).

Such a quantum ab-initio approach was initially performed by *Csizmadia et al.* (*90, 91*) and these authors concluded that $H_2N\text{–}PH_2$ has a *planar*-N structure, in disagreement with experiment. However, the method they employed was of such high quality that the discrepancy between theory and experiment had to be carefully analysed. When we closely examined the way in which *Csizmadia et al.* had performed the optimization: we found that, in order to save computing time, they had assumed during the whole calculation that the angles HNH and HNP were equal, as well as the angles HPH and HPN. We felt that this assumption, dubious from a chemical point of view, could be the cause of the above-mentioned discrepancy.

Upon optimizing the *full* set of geometrical parameters for $H_2N\text{–}PH_2$ within the CNDO approximation, we observed (*92*) that *Csizmadia*'s assumption actually leads to a *planar* N structure with HPH = HPN = 90° (in agreement with the 97° value provided by ab-initio calculation). If, on the other hand, the assumption in question

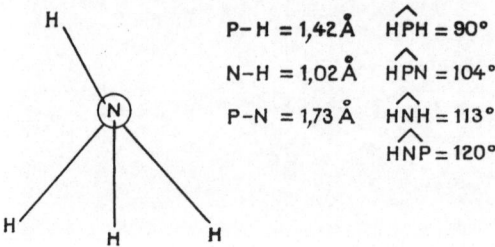

$$P-H = 1{,}42\,\text{Å} \qquad \widehat{HPH} = 90°$$
$$N-H = 1{,}02\,\text{Å} \qquad \widehat{HPN} = 104°$$
$$P-N = 1{,}73\,\text{Å} \qquad \widehat{HNH} = 113°$$
$$\widehat{HNP} = 120°$$

Fig. 26. Geometry and preferred conformation (Neumann projection) of $H_2N\text{-}PH_2$

is not included, one gets the optimized geometry and the preferred conformation given in Figure 26 in which the *pyramidal* character of the nitrogen atom is conspicuous, which is in complete agreement with experiment.

This example shows how careful one has to be with the results provided by a quantum technique, however sophisticated, when some ambiguous assumptions have been included in the calculations. It may even be preferable sometimes to use a more approximate method combined with assumptions made on chemical evidence.

IX. On the Nature of Conformational Parameters Provided by Experiment

The theoretical conformational analysis of a molecule, whatever the quantum technique used, provides quantities related to the free molecule at $0°K$ and within ideal standard entropy conditions. It follows that such results must be compared with experimental results obtained in conditions as close as possible to these. Obviously, any study in the gas phase will be preferable to corresponding ones performed on liquid or solid states. The most suitable experimental approaches will thus be electron diffraction and microwave spectroscopy.

From this point of view, let us wander a little from the subject to discuss briefly the comparison of the computed and experimental values of a dipole moment. Too often, people compare their theoretical results with experimental values obtained *in solution*, and if there is a discrepancy between the two sets, they generally blame the so-called failure of quantum chemistry to predict dipole moments.

From our own experience, we may assert that among the electronically localized systems we studied good agreement between theory and experiment was widely observed — at least with respect to the order of magnitude — *provided one compares what is comparable*, i.e. the CNDO computed values with the experimental ones obtained *in the gas phase* by means of microwave spectroscopy *using the Stark effect*. This agreement vanishes in many cases if the dipole moment is measured in solution. Such measurements in solution appear to depend strongly on solvation effects, which alter the μ_M value, hiding the molecular contribution under some first-order intermolecular perturbations. It would seem more convenient to use the comparison of computed ("in the gas phase") and experimental (in solution) data to indicate the importance of solvation than to test — in a basically misleading way — the quantum techniques used. This would be perhaps an indirect manner to approach any solute-solvent interaction quantitatively.

Coming back to the experimental methods which can be used for molecular conformational analysis, we include here a brief critical survey of a method much quoted in the literature, i.e. n.m.r. including the VT process. Reviewing our work over many years on phosphorus-nitrogen compounds, we compared the rotational barriers around the (P–N) bonds obtained by the coalescence n.m.r. procedure (*93*) with those obtained by microwave spectroscopy or electron diffraction. The main conclusion was that there is often an order of magnitude of difference between these two sets: for $(CH_3)_2N-PF_2$, for example, the barrier to internal rotation around the (P–N) bond is 2.5 kcal. mole^{-1} from electron diffraction measurements (*94*), 2.8 kcal. mole^{-1} from our calculations (*95*), and about 10 kcal. mole^{-1} as given by VTP n.m.r. (*93*).

In other words, it seems that when the coalescence procedure alone is used to determine a rotational barrier, i.e. to try to reach a ΔG^* value, it is actually a ΔH quantity which is obtained, the ΔG^* component being hidden by a huge $T \cdot \Delta S$ component, once more due to solvation factors. The n.m.r. approach might reveal the *trends* of a given barrier throughout a series of chemicals if ΔH paralleled ΔG^* in that case. Unfortunately, a critical analysis of ΔH (from n.m.r.) and ΔG^* (from microwave spectroscopy) data shows that the $T \cdot \Delta S$ contribution does not remain constant throughout many series, consequently introducing some crossings within the ΔH and ΔG^* scales. This difficulty has been taken into account during recent years by many n.m.r. specialists, who conduct their experiments so as to estimate not only ΔH but also $T \cdot \Delta S$, thus obtaining by difference ΔG^* values in excellent agreement with data from studies in the gas phase. The purpose of this paragraph is to put chemists on their guard against a careless use of n.m.r. within the field of conformational analysis.

X. Concluding Remarks

The purpose of this brief survey was to demonstrate that, despite the criticisms which may be made of the use of any semi-empirical quantum technique for structural and conformational studies, the CNDO/2 and Extended CNDO/2 formalisms are definitely reliable tools for theoretical conformational analyses in inorganic and coordination chemistry. Moreover, if these tools are combined with the most suitable experimental techniques (i.e. microwave spectroscopy and electron diffraction) in that field, many problems of geometry and conformation can be solved in a way that neither of these approaches could have accomplished alone.

The CNDO method appears to be from this point of view a *new spectroscopy* for the study of molecular conformations, provided that the molecules in question belong to the class of electronically localized systems.

This paper in defense of semi-empirical quantum techniques was written with the object of counterbalancing the excessive and incomprehensible dislike these methods nowadays arouse. *"IN MEDIO STAT VIRTUS"*, even in quantum chemistry!

References

1. *Kuczkowski, R. L., Lide, D. R.:* J. Chem. Phys. **46**, 357 (1967)
2. *Pople, J. A., Beveridge, D. L.:* Approximate Molecular Orbital Theory. New York: McGraw-Hill 1970
3. *Labarre, J.-F., Leibovici, C.:* Internat. J. Quantum Chem. **6**, 625 (1972)
4. *Pasinski, J. P., Kuczkowski, R. L.:* J. Chem. Phys., **54**, 1903 (1971)
5. *Labarre, J.-F., Leibovici, C.:* J. Chim. Phys., Fr. **69**, 404 (1972)
6. *Crasnier, F., Labarre, J.-F., Leibovici, C.:* J. Mol. Struct. **14**, 405 (1972)
7. *Bach-Chevaldonnet, M.-C., Crasnier, F., Labarre, J.-F., Leibovici, C.:* **20**, 141 (1974)
8. *Bryan, P. S., Kuczkowski, R. L.:* Inorg. Chem. **11**, 553 (1972)
9. *Gropen, O., Seip, H. M.:* Chem. Phys. Lett. **11**, 445 (1971)
10. *Perahia, D., Pullman, A.:* Chem. Lett. **19**, 73 (1973)
11. Cf. *Pople, J. A., Gordon, M. S.:* J. Amer. Chem. Soc. **89**, 4253 (1967); *Gordon, M. S.:* J. Amer. Chem. Soc. **91**, 3122 (1969)
12. *Malrieu, J.-P.:* J. Chim. Phys. Fr. **73**, 319 (1976); *Gregory, A. R., Paddon-Row, N. N.:* J. Amer. Chem. Soc., **98**, 7521 (1976)
13. *Bach, M.-C., Crasnier, F., Labarre, J.-F., Leibovici, C.:* J. Mol. Struct. **16**, 89 (1973)
14. *Robinet, G., Leibovici, C., Labarre, J.-F.:* Chem. Phys. Lett. **15**, 90 (1972)
15. *Corosine, M., Crasnier, F., Labarre, M.-C., Labarre, J.-F., Leibovici, C.:* Chem. Phys. Lett. **20**, 111 (1973)
16. *Bach, M.-C., Labarre, J.-F., Leibovici, C.:* J. Chim. Phys. Fr. **70**, 1181 (1973)
17. *Craig, D. P.:* Chem. Ind. 3 (1958) *Craig, D. P., Paddock, N. L.:* Nature **181**, 1052 (1958); *Craig, D. P., Hefferman, M. L., Mason, R., Paddock, N. L.:* J. Chem. Soc., 1376 (1961). *Heffernan, M. L., Mason, R., Paddock, N. L.:* J. Chem. Soc., 1376 (1961)
18. *Dewar, M. J. S., Lucken, E. A. C., Whitehead, M. A.:* J. Chem. Soc., 2423 (1960)
19. *Faucher, J.-P., Devanneaux, J., Leibovici, C., Labarre, J.-F.:* J. Mol. Struct. **10**, 439 (1971)
20. *Faucher, J.-P., Glemser, O., Labarre, J.-F., Shaw, R. A.:* C. R. Acad. Sci. (Paris) **279 C**, 441 (1974)
21. *Wiberg, K. A.:* Tetrahedron **24**, 1083 (1968)
22. *Crasnier, F., Labarre, J.-F.:* Topics in Current Chem. **24**, 33 (1971)
23. *Armstrong, D. R., Longmuir, G. H., Perkins, P. G.:* Chem. Commun. 464 (1972)
24. *Faucher, J.-P., Labarre, J. F.:* Phosphorus 3, 265 (1974)
25. *Cassoux, P., Labarre, J.-F., Glemser, O., Koch, W.:* J. Mol. Struct. **13**, 405 (1972)
26. *Pelissier, M., Labarre, J.-F., Vilkov, L. V., Golubinsky, A. V., Mastryukov, V. S.:* J. Chim. Phys., Fr. **71**, 702 (1974)
27. *Emerson, G. F., Ehrlich, K., Giering, W. P., Lauterbur, P. C.:* J. Amer. Chem. Soc. **88**, 3172 (1966)
28. *Almenningen, A., Haaland, A., Wahl, K.:* Acta Chem. Scand. **23**, 1145 (1969)
29. *Savariault, J.-M., Labarre, J.-F.:* Inorg. Chim. Acta **19**, L 53 (1976)
30. *Serafini, A., Savariault, J.-M., Cassoux, P., Labarre, J.-F.:* Theoret. Chim. Acta (Berlin) **36**, 241 (1975)
31. *Serafini, A., Pelissier, M., Savariault, J.-M., Cassoux, P., Labarre, J.-F.:* Theoret. Chim. Acta (Berlin) **39**, 229 (1975)
32. *Savariault, J.-M., Serafini, A., Pelissier, M., Cassoux, P.:* Theoret. Chim. Acta (Berlin) **42**, 155 (1976)
33. *Savariault, J.-M., Labarre, J.-F.:* Theoret. Chim. Acta (Berlin) **42**, 207 (1976)
34. *Finseth, D. H., Sourisseau, C., Miller, F. A.:* J. Phys. Chem. **80**, 1248 (1976)
35. *Fitzpatrick, N. J., Savariault, J.-M., Labarre, J.-F.:* J. Organometal. Chem. **127**, 325 (1977)
36. *Savariault, J.-M.:* unpublished results
37. *Brown, D. A., Fitzpatrick, N. J., Manning, A. R.:* J. Organometal. Chem. **102**, C 29 (1975)
38. *Almenningen, A., Jacobsen, G. G., Seip, H. M.:* Acta Chem. Scand. **23**, 685 (1969)

39. *Hargittai, M., Hargittai, I., Spiridonov, V. P., Pelissier, M., Labarre, J.-F.:* J. Mol. Struct. **24**, 27 (1975)
40. *Robinet, G., Leibovici, C., Labarre, J.-F.:* Theoret. Chim. Acta **26**, 257 (1972)
41. *Robinet, G., Crasnier, F., Labarre, J.-F.:* Theoret. Chim. Acta **25**, 259 (1972)
42. *Faucher, J.-P., Labarre, J.-F.:* Advances in Mol. Relax. Processes **8**, 169 (1976)
43. *Allcock, H. R., Bissell, E. C., Shawl, E. T.:* Inorg. Chem. **12**, 2963 (1973)
44. *Mani, N. V., Wagner, A. J.:* Acta Cryst. **B 27**, 51 (1971)
45. *Shaw, R. A.:* private communication
46. *McGandy, E. L.:* Ph. D. Dissertation, Boston Univ. (1961), Dissertation Abstr. **22**, 754 (1961)
47. *Robinet, G., Labarre, J.-F., Leibovici, C.:* Chem. Phys. Lett. **22**, 356 (1973)
48. *Graffeuil, M., Labarre, J.-F., Leibovici, C.:* J. Mol. Struct. **22**, 97 (1974)
49. *Graffeuil, M., Labarre, J.-F., Leibovici, C.:* J. Mol. Struct. **23**, 65 (1974)
50. *Graffeuil, M., Labarre, J.-F., Lappert, M. F., Leibovici, C., Stelzer, O.:* J. Chim. Phys. Fr. **72**, 799 (1975)
51. *Leibovici, C., Graffeuil, M., Labarre, J.-F.:* J. Chim. Phys. Fr. **72**, 272 (1975)
52. *Graffeuil, M.:* Thesis, Toulouse, France, 1976
53. *Jano, I.:* C. R. Acad. Sci. (Paris) **261**, 103 (1965)
54. *Drago, R. S., Wayland, B. B.:* J. Amer. Chem. Soc. **87**, 3571 (1965)
55. *Graffeuil, M., Labarre, J.-F.:* J. Chim. Phys. Fr. **73**, 1042 (1976)
56. *Anderson, G. A., Forgaard, F. R., Haaland, A.:* Chem. Commun. 480 (1971)
57. *Drago, R. S., Vogel, G. C., Needham, T. E.:* J. Amer. Chem. Soc. **93**, 6014 (1971)
58. *Thiele, K. H.:* Z. Anorg. Allg. Chem. **322**, 72 (1963)
59. Cf. *Graham, W. A. G., Stone, F. G. A.:* J. Inorg. Nucl. Chem. **3**, 164 (1956)
60. *Baldwin, R. A., Washburn, R. M.:* J. Org. Chem. **26**, 3549 (1961)
61. *Young, D. E., McAchran, G. E., Shore, S. G.:* J. Amer. Chem. Soc. **88**, 4490 (1966)
62. *Fridmann, S. A., Fehlner, T. P.:* Phys. Chem. **75**, 2711 (1971)
63. *Peterson, L. K., Wilson, C. L.:* Canad. J. Chem. **49**, 3171 (1971)
64. *Reetz, T., Katlafsky, B.:* J. Amer. Chem. Soc. **82**, 5036 (1960)
65. Cf. *Holmes, R. R., Carter, R. P.:* Inorg. Chem. **2**, 1146 (1963)
66. *Fleming, S., Parry, R. W.:* Inorg. Chem. **11**, 1 (1972)
67. *Jugie, G., Laussac, J.-P., Laurent, J.-P.:* J. Inorg. Nucl. Chem. **32**, 3455 (1970)
68. *Lunberg, K., Rowatt, R. J., Miller, N. E.:* Inorg. Chem. **8**, 1336 (1969)
69. *Jouany, C., Laurent, J.-P., Jugie, G.:* J. Chem. Soc. (Dalton) 1510 (1974)
70. *Durand, M., Jouany, C., Jugie, G., Elegant, L., Gal, J.-F.:* J. Chem. Soc. (Dalton) 57 (1977)
71. *Crasnier, F., Jouany, C., Jugie, G., Labarre, J.-F., Savariault, J.-M.:* J. Chim. Phys. Fr. **73**, 1036 (1976)
72. *Bartell, L. S., Brockway, L. O.:* J. Chem. Phys. **32**, 512 (1960)
73. *Lide, D. R., Mann, D. E.:* J. Chem. Phys. **28**, 572 (1958)
74. *Cassoux, P., Kuczkowski, R. L., Bryan, P. S., Taylor, R. C.:* Inorg. Chem. **14**, 126 (1975)
75. *Savariault, J.-M., Labarre, J.-F.:* unpublished results
76. *Almenningen, A., Andersen, B., Astrup, E. E.:* Acta Chem. Scand. **24**, 1579 (1970)
77. *Marriott, J. C., Salthouse, J. A., Ware, M. C., Freeman, J.:* Chem. Commun. 595 (1970)
78. *Kruck, T., Prasch, A.:* Angew. Chem. **356**, 118 (1968)
79. *Cassoux, P.:* unpublished results
80. *Chiu, N.-S., Schäfer, L., Seip, R.:* J. Organometal. Chem. **101**, 331 (1975)
81. Cf. *Stelzer, O., Schmutzler, R.:* J. Chem. Soc. 2867 (1971)
82. *Corosine, M., Crasnier, F., Labarre, J.-F., Labarre, M.-C., Leibovici, C.:* J. Mol. Struct. **22**, 257 (1974)
83. *Corosine, M., Crasnier, F.:* J. Mol. Struct. **27**, 105 (1975)
84. *Sheldrick, W. S., Stelzer, O.:* J. Chem. Soc. (Dalton) 926 (1973)
85. *Ayed, N., Mathis, R., Burgada, R., Mathis, F.:* C. R. Acad. Sci. (Paris) **278** C, 1085 (1974)

86. *Mathis, R., Ayed, N., Charbonnel, Y., Burgada, R.:* C. R. Acad. Sci. (Paris) **277** C, 493 (1973)
87. *Mathis, R., Burgada, R., Sanchez, M.:* Spectrochim. Acta **25** A, 1201 (1969)
88. *Labarre, M.-C., Coustures, Y.:* J. Chim. Phys. Fr. **70**, 534 (1973)
89. *Labarre, M.-C., Coustures, Y.:* C. R. Acad. Sci. (Paris) **276** C, 133 (1973)
90. *Csizmadia, I. G., Cowley, A. H., Taylor, M. W., Tel, L. M., Wolfe, S.:* Chem. Commun. *1147 (1972)*
91. *Csizmadia, I. G., Cowley, A. H., Taylor, M. W., Wolfe, S.:* Chem. Commun. 433 (1974)
92. *Barthelat, M., Mathis, R., Labarre, J.-F., Mathis, F.:* C. R. Acad. Sci. (Paris) **280** C, 645 (1975)
93. Cf. *Cowley, A. H., Dewar, M. J. S., Jackson, W. R., Jennings, W. B.:* J. Amer. Chem. Soc. **92**, 1085 (1970)
94. *Holywell, G. C., Rankin, D. W. H., Beagley, B., Freeman, J. M.:* J. Chem. Soc. A 785 (1971)
95. *Bach, M.-C., Brian, C., Labarre, J.-F., Leibovici, C., Dargelos, A.:* J. Mol. Struct. 17, 23 (1973)

The Approximate Calculation
of Molecular Electronic Structures as a
Theory of Valence

D. B. Cook

The Chemistry Department, The University, Sheffield S3 7HF, Great Britain

Table of Contents

Introduction

This paper is an attempt to 're-think' the problem of the approximate calculation of molecular electronic structure from the point of view of the theory of valence. That is, we regard the 'minimal basis atomic orbital expansion method' as a valuable conceptual aid in the understanding of molecular electronic structure and therefore worth investigating as fully as possible. In fact the understanding of valence is regarded here as an important problem *in its own right*. In this view it is, for example, more important to understand the physical processes occurring on bond formation than it is to be able to compute the corresponding bond energies accurately: the latter are accessible experimentally, the former are not. Thus we do not shrink from abandoning potentially accurate large scale ab initio CI or MCSCF methods *and* most of the existing semi-empirical techniques since both of these approaches (for different reasons) are largely uninterpretable as theories of valence (*1*).

No attempt is made to review the literature — references are only made to existing techniques as and when they impinge on the central ideas of this exposition. Occasionally we are critical of some of the established schemes: this is not, of course, to belittle the importance of the ideas of the pioneers in this field but rather to question the continued uncritical use of these ideas.

The *Variation Principle* is the main point of departure: all questions of symmetry, approximation etc. are judged from the point of view of their likely effect on the variational form of the Schrödinger equation. We attempt to take the minimal basis AO expansion method as far as possible while remaining within a family of well-defined conceptual models of the electronic structure which is theoretically and numerically underpinned by the variation principle.

The One-Configuration Model and the Variation Principle

We start our analysis from the co-ordinate independent (global) variation principle

$$\delta \tilde{E} = 0 \tag{1}$$

subject to the constraint

$$\delta \tilde{N} = 0 \tag{2}$$

where

$$\tilde{E} = \int dV \, \tilde{\Phi}^* \hat{H} \tilde{\Phi} \tag{3}$$

is the average value of the energy of a function and

$$\tilde{N} = \int dV \, \tilde{\Phi}^* \tilde{\Phi} \tag{4}$$

is the corresponding normalisation condition on $\tilde{\Phi}$.

If we admit *arbitrary* variations $\delta\tilde{\Phi}$ of $\tilde{\Phi}$ in (1) and (2) then

$$\delta\tilde{E} = 2 \int dV \, \tilde{\Phi}^* \hat{H} \, \delta\tilde{\Phi} = 0$$

and

$$\delta\tilde{N} = 2 \int dV \, \tilde{\Phi}^* \, \delta\tilde{\Phi} = 0 \, .$$

Hence

$$\int dV \, \delta\tilde{\Phi}^*(\hat{H} - W) \, \tilde{\Phi} = 0$$

for a Lagrange multiplier W and hence

$$\hat{H}\Phi = W\Phi \tag{5}$$

for the exact solution ($\Phi = \tilde{\Phi}$; $W = E$).

The transition from (1) and (2) to (5) is reversible; each implies the other *if the variations $\delta\tilde{\Phi}$ admitted are completely arbitrary*. More important from the point of view of approximation methods, Eq. (1) and (2) remain valid when the variations $\delta\tilde{\Phi}$ in a trial function Φ are *constrained* in some systematic way whereas the solution of (5) subject to model or numerical approximations is technically much more difficult to handle. By 'model approximation' we shall mean an approximation to the *form* of Φ as opposed to numerical approximations which are made at a lower level once a model approximation has been made. That is, we assume that $\hat{H}$, the molecular Hamiltonian is fixed (non-relativistic, Born-Oppenheimer approximation; which itself is a model in a wider sense) and we make models of the large scale electronic structure by choice of the form of Φ and then compute the detailed charge distributions, energetics etc. *within that model*.

This classification of approximations into 'model' and 'numerical' is not without ambiguities of interpretation. In particular we shall regard the use of an orbital basis as a model approximation. But calculations of molecular electronic structure are scarcely feasible without this approximation and so it could be regarded as a numerical convenience. However, in all that follows the energy integrals involving the orbital basis will be the prime candidates for numerical approximation and so we regard the orbital basis expansion technique as a model. The specific point 'modelled' by the finite basis expansion is that point properties are lost by the expansion technique which can only give a reasonable description of the electron distribution 'in the large': the variation principle is global not local. Within the confines of our use of the term 'model' — fixed $\hat{H}$ in the variational equation — we might therefore loosely differentiate between model and numerical approximations along the following lines. An approximation is a numerical approximation if, for some 'values' of that approximation the variation principle can be violated; conversely, an approximation is of a model type if all 'values' of the approximation 'obey' the variation principle. This formulation rather stretches the meaning of the word 'value' but the intention is clear. Thus, while the neglect of certain AO basis integrals may violate the variation principle the use of the MO model never will.

Since the Schrödinger equation for the system can be recovered from the variation principle by admitting arbitrary variations in a trial function Φ we expect to recover a Schrödinger-like[1]) equation if we constrain the form of Φ i.e. only admit variations $\delta\Phi$ of a certain restricted class. The models we make of molecular electronic structure should, of course, reflect as far as possible the qualitative concepts used to analyse that structure. The most pervasive concepts of chemistry are the ideas of localised bonds, functional groups and groups of delocalised electrons − separate electron groups. That is, the electronic structure of a large molecule is never seen conceptually as anything but a series of interacting separate electron groups with more or less well-defined separate identity and properties.

We are therefore led to make the *one-configuration separate electron group model* for molecular electronic structure. We write

$$\Phi = \hat{A} \prod_{R=1}^{N} \Phi_R \tag{6}$$

(dropping the tilde from Φ from now on) where Φ_R is a wavefunction for the 'R-group' of electrons (containing n_R electrons) and $\hat{A}$ is the total n-electron antisymmetrising operator. We make this approximation with the intention of identifying the Φ_R with the structure of (more or less) well-localised groups of electrons but (6) does not strictly require this. It is obvious that if we constrain the n_R to be all unity then we recover the simplest form of (6) − the MO model in which the Φ_R are delocalised (or can be chosen to be so). In order that (6) be most easily interpretable it is convenient to impose the Pauli principle *within* each group − i.e. we assume Φ_R is antisymmetric in its 'own' variables. Thus we set

$$\hat{A} = \hat{A}_X \prod_R \hat{A}_R \tag{7}$$

where $\hat{A}_R$ is the antisymmetriser for the variables of Φ_R alone and $\hat{A}_X$ contains only *inter-group* exchanges. The operator $\hat{A}_X$ can be expanded in terms of products of the operators $\hat{X}_{RS}$ which exchange variables between groups R and S only:

$$M \hat{A}_X = 1 - \sum_{R>S} \hat{X}_{RS} + \sum_{\substack{R>S \\ T>U}} \hat{X}_{RS} \hat{X}_{TU} - \dots \tag{8}$$

---

[1]) Clumsy nomenclature for an equation reminiscent of Schrödingers equation.

where

$$\hat{X}_{RS} = \hat{X}_{SR}$$

$$\hat{X}_{RS}\,\hat{X}_{TU} = \hat{X}_{TU}\,\hat{X}_{RS} \qquad (R,S \neq T,U)$$

and M is a normalising factor.

Therefore replacing Φ_R by $\hat{A}_R\,\Phi_R$ in (6) (without change of notation) we have

$$\Phi = \hat{A}_X \prod_{R=1}^{N} \Phi_R \,. \qquad (9)$$

Now $\hat{H}$ is of known structure and can always be expanded (if only formally) as:

$$\hat{H} = \sum_R \hat{H}_R + \sum_{R>S} \hat{V}_{RS} \qquad (10)$$

(where $\hat{H}_R$ only involves the variables of group R and $\hat{V}_{RS}$ only contains the variables of groups R and S) since $\hat{H}$ contains only one- and two-electron operators.

We now substitute (9) and (10) into (1) and (2). It is at once obvious that, unless we impose the so-called strong orthogonality constraint

$$\int dv_1\, \Phi_R{}^*(1,i\,...)\,\Phi_S(1,j\,...) = 0 \qquad (R \neq S) \qquad (11)$$

on our separate electron group functions Φ_R, the simplicity of our model is lost. Thus, in spite of choosing the form (9) for a trial wave function and the form (10) for a partition of the Hamiltonian we do not recover an expression

$$E = \sum_R E_R + \sum_{R<S} E_{RS}$$

for the total energy (with an obvious notation).

There are many extra terms in the energy expression arising from integrations after the operation of $\hat{A}_X$ *if* (11) does not hold. Since interpretation is one of our main aims we therefore unhesitatingly impose (11) on our functions Φ_R.

With this constraint we can expand the average value E of the energy of (9) over the Hamiltonian (10):

$$E = \int dV \hat{A}_X \prod \Phi_R{}^* (\sum_R \hat{H}_R + \sum_{R<S} \hat{V}_{RS})\,\hat{A}_X \prod \Phi_R \,. \qquad (12)$$

The action of $\hat{A}_X$ on *both* sides of (12) is merely to generate repetitions. Apart from a constant therefore, (12) becomes

41

$$\int dV \prod \Phi_R^* (\sum_R \hat{H}_R + \sum_{R<S} \hat{V}_{RS}) \hat{A}_X \prod \Phi_R$$

$$= \int dV \, \Phi_R^* (\sum_R \hat{H}_R) \hat{A}_X \, \Phi_R$$

$$+ \int dV \prod \Phi_R^* (\sum_{R<S} \hat{V}_{RS}) \hat{A}_X \prod \Phi_R \,. \tag{13}$$

Since $\hat{H}_R$ acts on only the variables of group R *and* Eq. (11) holds, the first term on the right of (13) becomes

$$\int dV \prod \Phi_R^* (\sum_R \hat{H}_R) \prod \Phi_R = \sum_R E_R \tag{14}$$

where

$$E_R = \int dV_R \, \Phi_R^* \hat{H}_R \, \Phi_R \,.$$

The second term of (13) can be evaluated with the help of (8) and the knowledge that the $\hat{V}_{RS}$ can only involve, at most, two-electron operators. If any $\hat{V}_{RS}$ contains only two-electron operators and the Φ_R are constrained by the strong orthogonality condition (11) then it is obvious that only the first two terms in the expansion (8) of $\hat{A}_X$ give rise to non-zero contributions to

$$\int dV \prod \Phi_R^* (\sum_{R<S} \hat{V}_{RS}) \hat{A}_X \prod \Phi_R \,. \tag{15}$$

Further, the only terms contributing to (15) are the ones contained in

$$\int dV \, \Phi_R^* \{ \sum_{R<S} \hat{V}_{RS} (1 - \hat{X}_{RS}) \} \, \Phi_R \tag{16}$$

because of strong orthogonality and the properties of the operators $\hat{X}_{RS}$. We therefore write a typical term of (16) as

$$V_{RS} = J_{RS} - K_{RS} \tag{17}$$

where

$$J_{RS} = \int dV_R \int dV_S \, \Phi_R^* \Phi_S^* \hat{V}_{RS} \, \Phi_R \, \Phi_S$$

and $\tag{18}$

$$K_{RS} = \int dV_R \int dV_S \, \Phi_R^* \Phi_S^* \hat{V}_{RS} \, \hat{X}_{RS} \, \Phi_R \, \Phi_S \,.$$

Thus the mean energy value for the one-configuration function (9), subject to (10) and (11) is

$$E = \sum_R E_R + \sum_{R<S} V_{RS} \qquad (19)$$

where E_R is given by (14) and V_{RS} by (17) and (18). This result is particularly satisfying since the (scalar) energy expression (19) has the same partitioned form as the (operator) expression (10) for $\hat{H}$: an obvious advantage in interpreting the terms in (19).

We are now in a position to apply the variation principle to optimise a function of the form (9). Using (19), (18) and (14) we find that the change in E due to a small variation $\delta\Phi_R$ in Φ_R is:

$$\delta E^{(R)} = 2 \{ \int dV_R \, \delta\Phi_R^* [\hat{H}_R + \sum_{S\neq R} \int dV_S \, \Phi_S^* \hat{V}_{RS}(1 - \hat{X}_{RS}) \, \Phi_S] \, \Phi_R \qquad (20)$$

where the variation in E has been written $\delta E^{(R)}$ to emphasise that the $\Phi_S(S \neq R)$ are kept constant. The variation $\delta N^{(R)}$ in the normalisation integral is easily seen to be

$$\delta N^{(R)} = 2 \int dV_R \, \delta\Phi_R^* \, \Phi_R . \qquad (21)$$

Thus combining (20) and (21) by the usual Lagrange multiplier method we have the non-linear differential equation

$$\{\hat{H}_R + \sum_{S\neq R} (\hat{J}_{RS} - \hat{K}_{RS})\} \, \Phi_R = E_R \, \Phi_R \qquad (R = 1, 2, ... N) \qquad (22)$$

where

$$\hat{J}_{RS} = \int dV_S \, \Phi_S^* \, V_{RS} \, \Phi_S$$

and $\qquad (23)$

$$\hat{K}_{RS} = \int dV_S \, \Phi_S^* \, V_{RS} \, X_{RS} \, \Phi_S$$

are inter-group 'Coulomb' and 'Exchange' operators respectively. This derivation is not rigorous — in particular some questions of the normalisation of the individual Φ_R have been skated over — but a more formal derivation gives the same equation. The derivation is general; it does not depend on the possibility of localising the groups of electrons, being simply constrained by the formal expressions (9) and (11.) We have, of course, derived (22) with an eye to interpreting the Φ_R as wave functions of 'chemically interpretable' groups of electrons.

Eq. (22) is therefore the Schrödinger-like equation for the motion of the electrons in group R. It is easy to see that in fact there are N equations (22) where N is the number of terms in (9). Further, these N equations are connected through the

$\hat{J}_{RS}$ and $\hat{K}_{RS}$ operators so that the solution of this system of N *simultaneous* partial differential equations must be iterative. The familiar SCF MO process is, of course, just the application of (22) when each group contains just one electron.

Eq. (22) have been derived from the *variation principle alone* (given the structure of $\hat{H}$): they contain only the single model approximation of Eq. (9): the typically chemical idea that the electronic structure of a complex many-electron system can be (quantitatively as well as qualitatively) understood in terms of the interactions among conceptually identifiable separate electron groups. In the discussion of the exact solutions of the Schrödinger equation for simple systems the operators which commute with the relevant $\hat{H}$ ('symmetries') play a central role. We therefore devote the next section to an examination of the effect of 'symmetry constraints' on the solutions of (22).

'Symmetry' Constraints on the Variational Solution

Formally, the set of operators, $\{\hat{G}_i\}$ say, which commute with $\hat{H}$ have the same eigenfunctions as $\hat{H}$, that is if

$$\hat{H}\Phi = E\Phi \quad \text{and} \quad \hat{G}_i\hat{H} = \hat{H}\hat{G}_i$$

then

$$\hat{G}_i\Phi = g\Phi$$

and these equations must have some consequences for our model of molecular electronic structure. It is clear for example if $\{\hat{G}_i\}$ is a set of operators defining the spatial symmetry of a molecule then Φ will exhibit this symmetry in the sense that the one-electron density

$$n \int dV_1' \, \Phi^* \, \Phi \qquad (dV_1' = dV_2 \, dV_3 \, ... \, dV_n)$$

will be the same at space-group equivalent points in the molecule. However it is equally clear that, if the Schrödinger equation is solved exactly then this property will come automatically out of the solutions Φ, even if we have no knowledge of the relevant point group. This is so because $\hat{H}$ contains a potential energy contribution which has the relevant pointgroup symmetry.

However, if we restrict the *form* of Φ by model approximations then we can no longer guarantee that the variations $\delta\Phi_R$ will be such as to maintain the symmetry of the total product wave function

$$\Phi = \hat{A}_X \prod \Phi_R . \tag{24}$$

Indeed it is easy to see that, in general, the symmetry of the model Φ will not be recovered by the variational solution since, if any *one* of the Φ_R departs from the symmetry of $\hat{H}$, then the 'coupling operators' $\hat{V}_{RS}$ will destroy the symmetry of the other Φ_S. Any departures from symmetry will quickly propogate throughout the Φ_R. We therefore expect that a model solution of the form (9) will have rather complicated behaviour in the variational process, for example each single-configuration approximation should show characteristic 'saddle point' behaviour when variations $\delta\Phi_R$ are admitted. The minimum in the variational expression when the $\delta\Phi_R$ are constrained to have the correct symmetry should also be a local maximum with respect to 'symmetry-breaking' variations $\delta\Phi_R$.

This argument is not restricted to spatial symmetry and in fact the most familiar example of the phenomenon is the Different Orbitals for Different Spins (DODS) technique for open electronic shells where the total spin function $\hat{S}^2$ takes the role of our $\hat{G}_i$ (in the one-electron-group model).

It is found empirically and of course is predictable theoretically that, when using a *model* for molecular electronic structure, the set of eigenfunction equations associated with the operators commuting with $\hat{H}$ are *constraints* on the action of the variation principle: if E_1 is computed from Φ_R subject to symmetry constraints and E_2 is computed *in the same model* with no such constraints then (2)

$$E_1 > E_2 .$$

Thus, there is an important decision of policy to be made at this point. Do we

(A) Simply solve the variational problem and hope to take care of the symmetry restrictions after the energy optimisation?

(B) Construct the Φ_R so that the total wave-function has the correct symmetry even if the individual Φ_R do not?

or

(C) Construct the Φ_R so that each function has the symmetry constraint applied?

Once stated in this way the advantages and disadvantages of each possibility are obvious. Clearly (A) has the advantage of being variationally the 'best' solution (lowest energy). Option (C), as we shall see shortly, can be a rather severe constraint in many important cases. The use of (B), while appearing an attractive compromise on general physical grounds, often takes us unwittingly outside the one-configuration model and has the unfortunate effect of enormously complicating technical implementations. If we wish to restore the spatial symmetry of a function computed by use of option (A) then, of course, we may well go outside the one-configuration model but this is *after* the variational process is complete and is a much smaller technical problem than (B). To look ahead slightly, we hope to solve the variational problem in a way which will avoid this final step.

In our discussion so far we have used a group of spatial symmetries as an easily-visualised example. It is possible to put all the 'symmetries' of the molecular Hamiltonian on the same formal footing by simply stressing the formal equivalence of all the operators commuting with $\hat{H}$. But this tends to obscure the very real differences of interpretation between the physical nature of these operators and their relation to $\hat{H}$. We therefore provisionally classify the symmetries into three types:

(i) 'Spatial' symmetries of $\hat{H}$ – for most molecules the finite point group of the rigid molecule.

(ii) Operators involving degrees of freedom not appearing in $\hat{H}$ because of its approximate form – principally 'spin'.

(iii) The Pauli Principle – a symmetry of the wave function which cannot appear in any Hamiltonian.

Of these three classes (ii) is most easily disposed of; clearly if a co-ordinate q_i does not appear in $\hat{H}$ then we can anticipate that the variation process will be completely indifferent to symmetry classifications involving q_i. Unless, of course, the form of the trial function is chosen with these variationally phantom degrees of freedom in mind. In the case of electron spin the unrestricted solution of Eq. (22) would not therefore lead to a total wave function which is an eigenfunction of operators depending on spin co-ordinates.

Symmetries of type (i) above – which would be augmented by the dynamical symmetry groups of molecules if these were known and usable – can be ignored with impunity in the sense that the computed total energy is still bounded from below by the true energy. The Pauli principle is a unique type, it cannot be ignored either at the model level or at the level of seeking variational solutions of the full Schrödinger equation. Thus, unless the antisymmetrising operator is applied, the variational procedure breaks down completely. We have implicitly recognised the sui generis character of the Pauli principle in choosing our one-configuration model as

$$\Phi = \hat{A}_X \Pi \Phi_R .$$

It is sobering to consider the effect of the various kinds of symmetry constraint on models of molecular electronic structure. Taking the variationally necessary antisymmetry requirement first; one consequence of the Pauli principle on an exact solution of a many-electron Schrödinger equation is that the probability of two electrons occupying the same point in space (with the same spin) is zero. Now this is scarcely an irksome constraint on the motion of two particles which, in any case, repel each other strongly. However, if we go to the bottom of our family of models and take the closed-shell MO method then we find that the antisymmetry requirement is much more severe. In this model electrons of the same spin are forced to occupy different orthogonal *orbitals* – that is an electron is excluded from a whole region of space. Similar considerations apply to the spatial symmetries. Obviously the electron density function generated by the exact wave function has the spatial symmetry of the molecule: the electron density is the same at point-group equivalent points in space. If this symmetry constraint is applied to the simple MO model (in its strongest form, (C) above) then the individual MO electron density functions – the squares of the MOs – are constrained to be the same at equivalent points in the molecule. That is the electron density due to the whole system of n electrons is (in the MO model) composed of n separate contributions each of which has itself the molecular symmetry. This seems intuitively unreasonable to say the least. To ask that the symmetry constraints be satisfied at the level of the Φ_R seems to be extremely heavy-handed and may well prove very restrictive on the variational solution of the Eq. (22).

It is not unreasonable to make a general statement – the application of 'symmetry' constraints to a model wave function is more and more restrictive the simpler the model. In particular, application of symmetry restrictions at level (C) is courting disaster when very simple models are used.

Everything we have said in this section applies to the *integro-differential* Eq. (22). We therefore now examine the consequences of the transition to an orbital basis expansion method.

Consequences for the Orbital Expansion Method

The simple orbital basis expansion method which is used in the implementation of most models of molecular electronic structure consists of expanding each Φ_R as a linear combination of determinants of a set of (usually) atom-centred functions of one or two standard forms. In particular most qualitative and semi-quantitative theories restrict the terms in this expansion to consist of the (approximate) *occupied atomic orbitals* of the constituent atoms of the molecule. There are two types of symmetry constraint implicit in this technique.

(a) If identical atoms are to have identical orbitals centred on them then we are almost automatically using constraint option (C) above.

and

(b) Usually the atomic orbitals are much more symmetrical than the molecular environment requires – for example the use of the same 2 p AO for both σ and π orbitals of HF is a rather severe constraint on a simple LCAO MO model, the variation principle cannot act on the π AO.

Now any decision to break from these simple symmetry ideas must not be taken lightly since, as we shall see, this step automatically excludes the use of a whole class of semi-empirical methods for the evaluation of molecular integrals.

The very simplicity of the simple (minimal basis) AO expansion technique arises from the fact that, of all the processes occurring on molecule formation, only the *inter-atomic* electronic reorganisation is given recognition by this method. Orbital

47

contraction or expansion and anisotropic re-distributions of electron density around atomic centres are scarcely accessible to the minimal-basis AO expansion method without a prohibitively large increase in the 'atomic' orbital basis. Also, once AO contraction/expansion effects are admitted the problem of the choice of molecular-optimised atomic orbitals occurs and several non-linear degrees of freedom are added to the variational problem with associated technical difficulties. The attraction of the fixed-basis AO expansion technique consists of the fact that it is technically easy to implement – all optimisations reduce ultimately to the linear variation method: matrix diagonalisation.

We see therefore that, however desirable it is to abandon symmetry constraints from the point of view of the variation method, we shall be involved in radical departures from the conventional AO expansion method – particularly in its minimal basis form. Indeed most of the changes required to the usual AO basis method are already implicit in any decision to lower the symmetry of the AO basis from its local spherically symmetric form to that of the molecular point group. We shall see later that these considerations are too pessimistic.

Other Constraints on the Exact Wave Function

There are a number of very general theorems concerning the properties of the exact solutions to a given Schrödinger equation some of which have been used as constraints on approximation techniques. The most widely-known of these relationships is the Virial theorem (and its relatives the various hyper-virial theorems). The virial theorem states that, for a Hamiltonian composed of kinetic energy terms and Coulomb potentials, the total energy of the system is minus the kinetic energy ($E = -T$). In quantum mechanics the theorem relates the energy eigenvalue and the average value of the kinetic energy, in classical mechanics the related quantities are the total energy and the time average of the kinetic energy. We naturally ask whether it is appropriate to *impose* this condition on our model solution of the variational problem.

In contrast to the symmetry requirements, the virial theorem is a *dynamical requirement* and, with the exception of atoms, can only be tested once the solution of the variational problem has been carried through. Or, to be a little more cautious, the imposition of the virial theorem on the *form* of a model of the molecular electronic structure is not easy. (It should be said at this point that the simple form, $E = -T$, of the virial theorem for molecules is only valid *at equilibrium*; for the moment we simply pass over the problem of the choice between computed and observed molecular conformation).

There is, of course, a whole range of computed mean values of physical quantities which, for the exact wave function, should be equal to the corresponding observed molecular property.[2]) But the reproduction of these quantities is not different in kind from the calculation of the energy (apart from the existence of the variation principle) and it is, therefore, straining the concept of 'constraint' a little to reproduce (say) the exact molecular dipole moment. If quantities like dipole moment are reproduced exactly by constraint it is difficult to say in what sense we are *computing* the molecular electronic structure and electron distribution.

It is not difficult to anticipate the remaining problem – if one decides to impose some or all of the constraints which we have discussed (and assuming that this is technically possible!), are they *consistent*? It is scarcely conceivable that the consistency (or otherwise) of the various constraints we have discussed will be *independent of model*; that is capable of being satisfied by *any* choice of the one-configuration separate electron group model. This matter is difficult to consider in general and consideration of a simple example will expose these difficulties clearly.

Constraints on the Variation Principle: an Example

The minimal basis calculation on the hydrogen molecule is a well-worn but eminently suitable example for our purposes. It has a convenient symmetry element and orbital basis calculations can be carried through which are quantitatively acceptable and yet not prohibitively unwiedly to report. We give below variational calculations on the H_2 molecule using the familiar simplest AO basis in the one-electron-group (MO) model and the electron-pair (VB) model. These calculations have been performed explicitly to investigate the effect of 'symmetry constraints'.

First, a preliminary look at the spatial symmetry constraint; using a basis of two 1 s AOs the only symmetry at our disposal is the reflection perpendicular to the internuclear axis: interchanging the basis orbitals. Using the separate-atom 1 s orbitals (STOs, exponents 1.0), the closed shell LCAOSCFMO wavefunction gives a total energy of -1.090942 a.u. at the equilibrium internuclear separation of 1.4 a.u. Fixing one of the AOs as the separate-atom function and optimising the other by optimising the non-linear scale factor (orbital exponent) gives an energy of -1.119332 a.u. for optimum exponent 1.375. The values of the virial ratio $(-E/T)$ are, for the two functions, 1.336639 and 1.085062 respectively. Clearly then, for arbitrarily chosen basis orbitals in the linear variation method there seems to be a 'symmetry dilemma' (2) insofar as the lowest energy wave function does not have the molecular symmetry: it is possible to obtain an energy lower than the symmetry-restricted separate-atom basis by allowing the orbital basis to vary. But, when we allow both

---

[2]) Disregarding relativistic and other effects excluded from $\hat{H}$.

orbital exponents to vary independently to minimise the energy we find a total energy of -1.128181 a.u. for both orbital exponents *equal* at 1.190, virial ratio 1.006324. Thus there is (at least for our example) an AO basis which is variationally optimum *and* has the symmetry of the molecule.

These results are confirmed by the corresponding VB calculations using the full 'CI' of three singlets from two orbitals (Heitler-London plus two ionic structures). The separate-atom AO basis gives -1.10388 a.u.; the single optimised exponent gives -1.13463 a.u. (exponents 1.0 and 1.333) and the completely optimised basis -1.14518 (both exponents 1.201). We shall return to this point later since, as we have established then, these conclusions are only valid at one value of the internuclear separation (the experimental value, 1.4 a.u.).

The virial ratio is, as we noted above, 1.3366 for the separate-atom AO basis MO calculation, i.e. not 1.0. Now within the confines of the *linear* variation method (the usual LCAO approach) there is no remaining degree of freedom to use in order to constrain the virial ratio to its formally correct value (or indeed to impose any other constraint). Thus imposing the correct virial ratio on the linear variation method is, in this case, not possible without simultaneously destroying the symmetry of the wave function. Only by optimising the non-linear parameters can we improve the virial ratio as the above results show. Even at this most elementary level, the imposition of various formally 'correct' constraints on the wave function is seem to generate contradictions.

The conclusion above that optimisation of the non-linear parameters in the AO basis leads to a basis with correct spatial symmetry properties cannot be true for all internuclear separations. At $R = 0$ the orbital basis must pass over into the 'double-zeta' basis for helium i.e. two *different* 1 s orbital exponents. It would be astonishing if this transition were discontinuous at $R = 0$. While considering the variation of basis with internuclear distance it is worth remembering that the closed-shell spin-eigenfunction MO method does not describe the molecule at all well for large values of R: the spin-eigenfunction constraint of 'two electrons per spatial orbital' is completely unrealistic at large internuclear separation. With these facts in mind we have therefore computed the optimum orbital exponents as a function of R for three wave functions:

  (i) The independent-electron model with no spin or spatial constraints (UHF)
 (ii) The conventional closed-shell MO method (RHF)
(iii) The electron-pair model (VB, as described above).

In all cases both orbital exponents were allowed to vary independently.

The results of these calculations are summarised in Figs. 1, 2 and 3. Fig. 1 has plots of the total energy for the three wave functions (together with the results of *Kolos* and *Roothaan* (8) for reference). The optimised orbital exponents are plotted in Fig. 2. In Fig. 3 the orbital exponents for case (iii) for the short internuclear distance region: $R = 0$ to $R = 0.5$ a.u.

Taking Figs. 1 and 2 first, there are two very obvious points to make. Most striking are the qualitative difference between the RHF curve and the other two; the RHF

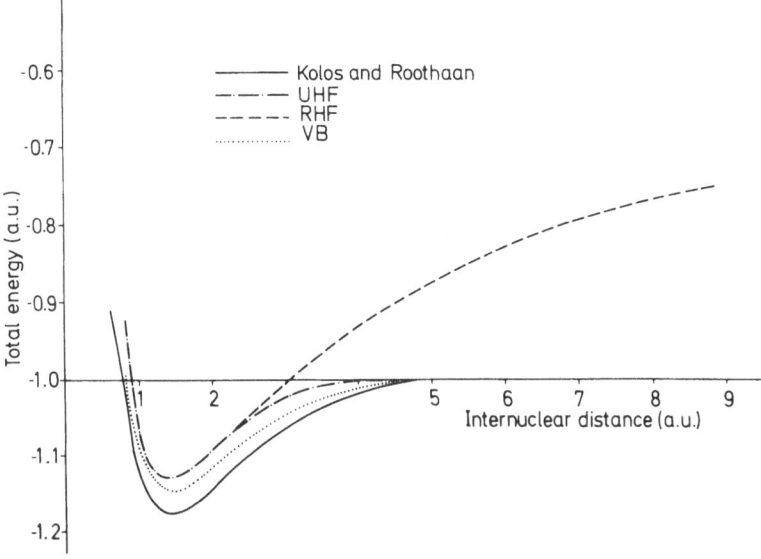

Fig. 1. The Hydrogen Molecule: total energy as a function of internuclear separation, RHF, UHF and VB

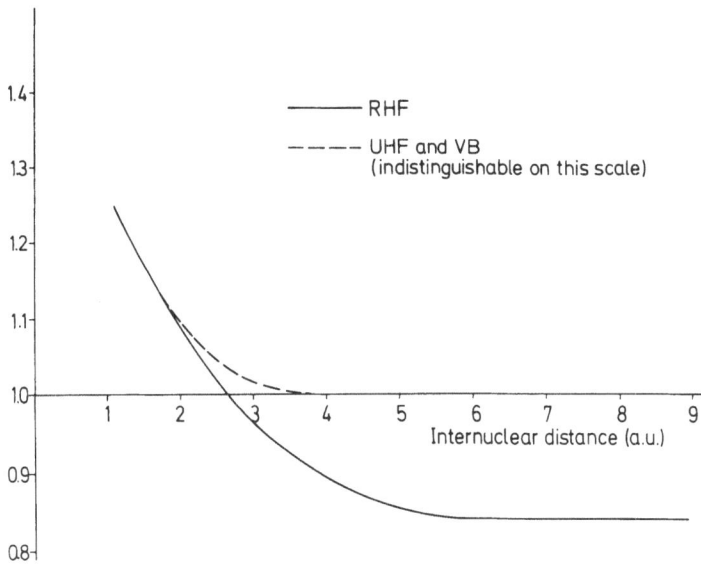

Fig. 2. The Hydrogen Molecule: optimised orbital exponents for the functions of Fig. 1

curve in Fig. 1 clearly goes to an incorrect limit as $R \to \infty$ and, more important, starts to diverge from the UHF curve at values of R quite close to equilibrium; $R \sim 2.0$ a.u. (~ 1.0 Å) is the separation point[3]). Secondly, the rather more realistic form of the UHF curve is noticeable: it coincides with the RHF curve around equilibrium and 'dissociates' to the separate atoms correctly. Thirdly, the well-known superiority of the electron-pair (VB) model is apparent from the figure. Finally, it is worth noting that, although there are small differences of a few per cent, the optimum orbital exponents for the UHF and VB wave functions as plotted in Fig. 2 are, by and large, identical. That is the optimum orbital basis is largely *independent of model* provided the model is physically realistic (i.e. excluding the RHF model away from equilibrium internuclear distance). Also, of course, there is only *one* orbital exponent quoted: in the region covered by Figs. 1 and 2 the optimised exponents are equal; the AO basis has the symmetry of the molecule.

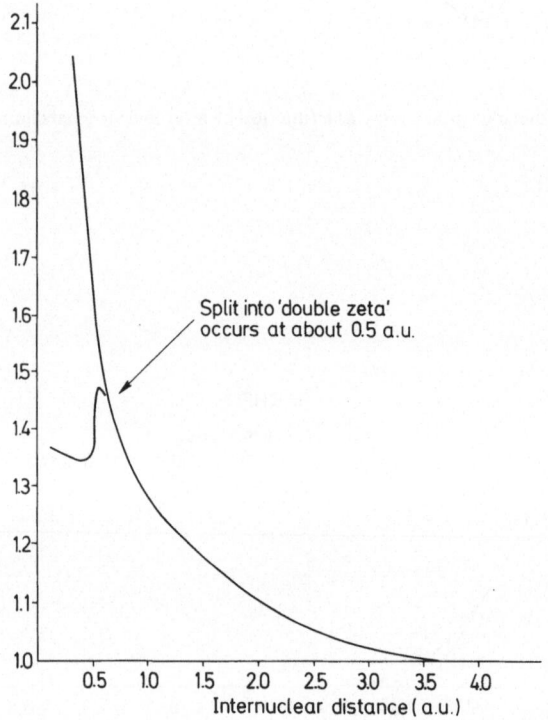

Fig. 3. The Hydrogen Molecule: optimised orbital exponents for small internuclear separation

---

[3]) At $R = \infty$ the total energy of an MO wave function constructed from two 1 s AOs (exponent δ) is $E = \delta^2 - 2\delta + 5/16\delta$ which is optimum for $\delta = 27/32$, ($E = -0.7119$ a.u.) clearly not the separate-atom AOs.

On the strength of these results alone there seems little justification for use of the RHF method for the calculation of molecular electronic structure (except for the technical consideration of time saving). The UHF method is identical with the RHF where the latter is at its best and has correct qualitative behaviour where the RHF method breaks down. The behaviour of the RHF energy even at moderate internuclear separations cannot be relied upon. The imposition of the spin eigenfunction constraint is disasterous for the MO function: to sacrifice one third of the total energy *and* lose the qualitative behaviour of the energy curve at moderate and large values of R for the sake of a classification of states involving variables *not appearing in the Hamiltonian* is, to put the point no more strongly, eccentric. Speaking a little more objectively, constraining the molecular wave function to be an eigenfunction of variables not appearing explicitly in the Hamiltonian involves both an unacceptably severe restriction on the action of the variation principle *and* a description of the physics of the system which is qualitatively wrong. The 'two electrons per orbital' rule is only appropriate where the electrons are constrained by the potential to 'share the same space'; at larger internuclear separations the concept is forced and artificial. Perhaps in this context it is useful to regard the UHF method as *the* independent-electron model and the RHF method as a degenerate form of the electron-pair model with two electrons per spatial orbital.[4])

The electron-pair curve is qualitatively correct at all values of R and is quantitatively better than either RHF or UHF. At first sight therefore, the VB method looks infinitely preferable to the RHF method and quantitatively superior to the UHF method. But, using only *one* electron pair the effect of the strong orthogonality constraint *between* electron pairs is disguised (there are no other electron pairs to which the bond-pair must be orthogonal). We shall examine the effect of the strong orthogonality constraint in a later section; for the moment we simply note the good numerical performance of the electron-pair model.

Now, for completeness, we turn to an examination of the optimised AO basis at short internuclear distance. We see in Fig. 3 that below $R \sim 0.5$ a.u. the optimised 1 s AO exponents on each nucleus of H_2 do differ — reflecting the $R = 0$ limit, the helium double-zeta basis. Thus, the 'symmetry dilemma' seems to make an appearance here. But, although the difference between the AOs does strictly violate the molecular symmetry, there is a more prosaic feature underlying the optimum basis. At small internuclear distances the overlap between the two AOs is very large if they are constrained to be the same. That is, the basis which has the molecular symmetry becomes increasingly redundant as R approaches zero. Thus, if we were to retain a basis with molecular symmetry we would be effectively using only one orbital and so lose one variational parameter. The 'splitting' of the orbitals reflects this variational process rather than something more fundamental. This conjecture is easily verified by using an orbital basis of a different functional form which does not have the same overlap behaviour with varying R. Putting considerations of the physical realism of the basis

---

[4]) The use of the UHF method as the 'standard' for the independent electron model also rationalises the definition of correlation energy (*9*).

aside and concentrating on the symmetry, an obvious choice for such a test is a basis which has overlap tending to zero as the internuclear distance decreases. A basis of one sp hybrid on each centre fulfills this requirement; at $R = 0$ the overlap is zero as it is at $R = \infty$. There is a maximum in the overlap at R around equilibrium. Carrying through the calculations does, in fact, confirm the analysis given above. At small values of R the optimum sp basis has two equal orbital exponents (including $R = 0$. At moderate values of $R(1.0-6.0)$ the optimum exponents differ and at large R, the exponents are again identical. Thus the optimised exponents mirror the redundancy in the orbital basis (the size of the overlap intehral). These calculations were all carried through using the VB model.

The conclusions from this rather elementary survey of the 'symmetry constraint' problem all point in the same general direction. The imposition of 'symmetry' constraints (other than the Pauli principle) on a variationally-based model is either unnecessary or harmful. Far from being necessary to ensure the physical reality of the wave function, these constraints often lead to absurd results or numerical instabilities in the implementation. The spin eigenfunction constraint is only realistic when the electrons are in close proximity and in such cases comes out of the UHF calculation automatically. The imposition of molecular spatial symmetry on the AO basis is not necessary if that basis has been chosen carefully – i.e. is near optimum. Further, any 'breakdowns' in the spatial symmetry of the AO basis are a useful indication that the basis has been chosen badly or is redundant.

As we stated at the outset, the variation principle is our starting point and the interpretation of valence and molecular electronic structure our aim: the imposition of some of the more familiar constraints on the action of the variation principle deflects us from that aim.

Numerical Approximations

Having chosen to work within a particular one of our one-configuration family of models, there are some important decisions to be taken about specific computations within that model. Most important of these is the question of *numerical approximation*: whether numerical approximations *are* to be used and, if so, how.

If we decide to use some form of orbital basis expansion method then one important step in the calculation is the generation of the various molecular integrals – the integrals involving orbital products and the terms in the Hamiltonian. The ab initio calculation of these numbers is a lengthy and galling task: lengthy because there are so many of them and galling because so many of them turn out to be very small. Thus we look to these integrals for the possibility of numerical approximation. If this is to be our chief area of numerical approximation we must now decide what role the numerical approximations will play. Will they be used to

(i) Approximate as closely as possible the largest (retained) molecular integrals in order to reproduce the full model calculation,

or

(ii) Attempt to correct or compensate for the deficiencies in the model approximations.

Option (i) leaves us with a well-defined technical problem which requires numerical solution (albeit guided by a knowledge of the physical meaning of the integrals). Option (ii) generates a curiously ill-defined hybrid problem to which there is no satisfactory answer. For example, it is known on theoretical grounds that the one electron group MO model includes no electron-electron correlation: each electron only experiences the averaged field of the others. Should therefore the approximate values of the molecular integrals used in such an MO theory be adjusted to 'allow for correlation effects' — typically made smaller than the corresponding exact values. This seems an intuitively reasonable suggestion, particularly if the MO model persistently gives poor numerical results. Making this choice for the use of numerical approximations should not be done lightly, however, since it constitutes an extremely restrictive condition on the model approximation. For example using such 'corrected' values of molecular integrals, while perhaps improving the agreement between computed and observed total energy, destroys the validity of the model approximation and so makes the results of such calculations uninterpretable as theories of valence. That is not to say that some of the existing parametrisations of the MO method are not valuable statistical tools for the extrapolation and interpolation of molecular energies, it is simply that these parametrisations are not theories of molecular electronic structure. What, for example, is the scientific value of a configuration interaction calculation (which includes some electron correlation in the model) using molecular integrals which 'contain electron correlation'? How many times has electron correlation been included in such a calculation; once, twice, one and a half times? If the numerical and model approximations are not kept separate then the interpretative value of the model is lost. Since our prime aim is an interpretable theory — a theory of valence — we judge all numerical approximations by their adherence to option (i) above.

In examining numerical approximations it is as well to bear in mind the general qualitative conclusion of our brief examination of symmetry constraints. In broad terms the result was: the simpler the model the more severe the effect of any constraint on the variation principle. This result cannot be carried over directly and used in numerical work since numerical approximation schemes can rarely be brought into a sufficiently coherent logical and mathematical form for analysis. Nevertheless it seems likely that this result can be used as a guideline — a 'rule of thumb'. We therefore expect that the imposition of formal constraints and consistency requirements (derived from a higher level of approximation or the exact solution) on numerical approximation schemes is likely to have far-reaching consequences — particularly on the simpler models of electronic structure. As we shall see, the imposition of apparently innocuous 'principles' at the level of a numerical approximation scheme may well render the approximated integrals meaningless in a more general model.

Before attempting to see how the problems of numerical approximation might be attacked we review briefly two of the more familiar 'consistency requirements' in approximate MO theory.

The 'invariance principle' of Pople and Segal (3) has been discussed elsewhere (4). Briefly, the use of neglect of differential overlap (NDO) approximations makes the MO method basis dependent in the sense that the result of an NDOLCAOMO calculation is not independent of linear transformations among the AO basis. This is a result with an extremely obvious physical interpretation; it is merely a re-statement of the fact that some AO bases will be better adapted to the neglect of 'many-centre' electron repulsion integrals than others. It is a matter of empirical investigation to find an optimum basis for NDO methods – orthogonalised hybrid AOs have been found to perform very well in this area (5). Pople and Segal, however, chose to impose the condition of invariance against linear transformations on the NDO approximate scheme, which has a series of ramifications. First, the usual physical interpretation of the electron repulsion integrals is lost since the directional characteristics of the non-s-type orbitals are abandoned. Thus it is, for example, difficult to know what the value of a localisation procedure is when the electron repulsion integrals are unaffected by many linear transformations.

Secondly, the invariance condition on the exact MO wave function is peculiar to the single determinant model[5]). If, therefore, the invariance condition is imposed on an approximation scheme and this imposition generates constraints on the *values* of molecule integrals, then the 'molecular integrals' obtained in this way can only at best have meaning *inside the MO model*. There is no valid possibility of using these integrals in a more general model of molecular electronic structure. Third, since the use of an invariance requirement of this type generates invalid equivalences among the molecular integrals *and* these integrals are calibrated against some standard results (experiment or ab initio) then the defects generated by insisting on an invariance principle must be made good by the *values* of the parameters – an unfortunate mix-up of model and numerical approximations. Finally, in the original proposals (3) only certain classes of linear transformation were considered: consideration of the full class of linear transformations leads to absurd results (4).

The second of these four consequences has proved to be the most unfortunate. Even when a set of parameters has been consciously optimised within the MO model (and there can be no objection of principle to the conscious use of the MO framework as a numerical interpolation device), the temptation to 'improve' on the MO results has proved irresistible. We can therefore find CI and VB calculations using molecular integrals which have been constrained by the invariance requirement to be meaningful only in the MO framework.

---

[5]) Excluding the complete CI expansion, of course.

Lindenberg (6) has proposed a novel and formally neat scheme for the non-empirical calculation of the bonding parameters (the β's) involved in the approximate MO theory of conjugated π systems. This elegant piece of work provides a good example of the effect of 'correct' consistency requirements and so we evaluate it carefully. The central idea is to use the formal relationship

$$[\hat{r}, \hat{H}] = i\,\hat{p}/m \tag{25}$$

(the equation of motion for the position operator in Heisenberg form). Linderberg shows that this relationship together with a traditional approximation for the dipole moment operator generates the formula

$$\beta_{rs} = \frac{1}{mR}\frac{dS_{rs}}{dR} \qquad (R = |R_r - R_s|) \tag{26}$$

for the π-electron β's (where ∇S has no out-of-plane component). As is noted in (6) the constraint that $S = 0$ does not necessarily imply $dS/dR = 0$, i.e. the equation for β_{rs} seems consistent with an orthogonal basis. It is worth noting here that these β's have no potential energy contribution — they are independent of core potential, a point which we will pursue later. It is our contention that, although on an impeccable theoretical foundation, the use of a formal equation to enforce consistency on a set of empirical approximations is out of place on three counts. Firstly, the general arguments given above suggest that constraints based on relationships satisfied by the exact wave function are likely to be unduly restrictive on approximation schemes. Secondly, in a certain sense, formal inconsistency is at the heart of numerical approximation methods: it is hardly consistent to set $(ij, ij) = 0$ when (ii, jj) differs appreciably from its asymptotic value of $1/R$, for example. In particular, to constrain the results of a variational calculation to be consistent with one of its *possible* uses (involving operators not appearing in $\hat{H}$) would paralyse research. As we have seen these type of constraints may well not be consistent among themselves. Finally, to emphasise the 'double-edged' nature of consistency requirements, the use of dS/dR for an orthogonalised basis is to misunderstand the nature of the orbital basis which validates 'neglect of differential overlap' theories. The basis is not simply orthogonal at some particular molecular geometry and non-orthogonal elsewhere, it is an orthogonal*ised* basis. That is, as the geometry changes *the orbital basis changes* so as to maintain orthogonality. Thus $dS/dR = 0$ since $S(R)$ is constrained to be zero. Thus this technique is inconsistent with the very foundations of the ZDO method.

We therefore conclude that attempts to impose constraints which are based on the formal properties of the exact Schrödinger equation leads to contradictory and even self-contradictory results besides placing unnecessary limitations on the action of the variation principle. That is not to advocate wilful inconsistency, of course. We shall insist on consistencies but within the confines of our orbital-basis variational model. In particular we shall use the following guidelines for numerical approximation within the orbital basis method.

(i) We use an orbital basis which is well-adapted to the particular numerical approximation scheme being used — using, for example, transformations among the basis functions etc. to validate our chosen numerical approximation method. There is the *possibility* of conflict here with an orbital basis chosen on intuitive chemical grounds but in practice we have never found such a conflict — quite the contrary.

(ii) We aim to reproduce the value of each type of molecular integral over the chosen basis as closely as possible: either by explicit calculation or by some approximation technique. Thus, we may choose in our approximation scheme simply to omit certain (presumed small) integrals and compute the others explicitly or we may approximate the molecular integrals retained or any other justifiable scheme.

These 'rules' certainly seem innocuous enough but they do represent an approach to the approximate calculation of molecular electronic structure which is fundamentally different from the bulk of the existing methods loosely called 'neglect of differential overlap' or 'semi-empirical'. The most important departure is in (ii) above. Our molecular integrals *will not be parameters*: their sole reason for existence is to be as close as possible in value to the exact integral. That is, we do not rely on incorporation of experimental data, cancellation of errors or any other divine dispensations. If our approximations generate poor results (in the sense of poor agreement with experiment) then these 'poor' results are just as valuable as 'good' results: they enable us to *evaluate and refine* our approximations and/or our model *without the loss of the interpretability of our model*. We will not, for example, be obliged to force agreement between an MO model and an unwilling nature; we can use our molecular integrals in a more general model of electronic structure. We are thus placing the major emphasis is again on the interpretation of results and taking the general view that quantitative agreement with experimental energy values, while important, is a *source of evidence* for the evaluation of theories of electronic structure (theories of valence) not an end in itself. In particular *pre-arranged* agreement with experiment is of no theoretical value.

We now turn to an examination of some of the necessary constraints on an orbital-basis implementation of our family of models for molecular electronic structure.

The Strong Orthogonality Constraint

In the first section of this work, in order to obtain maximum simplicity of interpretation, we chose to impose the strong orthogonality constraint on our model wave functions: any two separate-group functions will be constrained by Eq. (11):

$$\int dv_1\, \Phi_R^*(1, i, ...)\, \Phi_S(1, j, ...) = 0 \qquad (R \neq S).$$

Obviously, this constraint has some consequences for the orbital expansion technique – it restricts our freedom in choosing an orbital basis. In fact, as Löwdin (7) has shown, in the context of the orbital expansion method (11) is equivalent to the constraint that each orbital used in the construction of Φ_R must be orthogonal to all the orbitals used to form Φ_S. Clearly then, the strong orthogonality requirement is model-dependent since different models of a given molecular electronic structure may correspond to different partitionings of the orbital basis and hence to a different orthogonality conditions among the orbitals of the basis set.

In this section we examine this orthogonality constraint in order to evaluate its consequences for a theory of valence. Is it a substantive formal constraint on the type of model we may use; does it restrict the type of physical phenomenon we can describe or is it simply a technical constraint on the method of calculation or what? In fact we shall find that the strong orthogonality constraint is central to any orbital basis theory of molecular electronic structure. It has a bearing on the applicability of the model approximations we use, on the validity of most numerical approximations used within these models and (apart from the simplest MO model) has a dominant effect on the technical feasibility of the methods of solution of the equations generated by our models. Thus, it is of some importance to try to separate these various effects and attempt to evaluate them individually.

In practice it is most common and convenient to side-step one of the problems associated with strong orthogonality. We can work with an orbital basis which satisfies (11) *independently* of the choice of the physical structure of the groups of electrons: a basis for which (11) is guaranteed for *all* Φ_R, Φ_S. Any orbital basis which forms an orthogonal set will fulfill this condition; all overlap integrals are then zero.[6]

$$S_{ij} = \int dV \, \bar\phi_i \, \bar\phi_j$$

We use an upper 'bar' to distinguish the members of an orthonormal set; thus $\bar\phi_i, \bar\phi_j$. This choice simplifies the comparisons between different models; it enables the use of the same basis and therefore the same integrals in the variational calculation. We now take up the three areas of implementation, numerical approximation and model approximations separately.

We have already mentioned that the question of technical implementation is one of the main reasons for the use of the orthogonality constraint. If this constraint is not imposed the huge number of terms in the energy expression (the n! problem) effectively prohibits quantitative calculations on any but the smallest molecular systems. The one-electron-group MO method is an exception to this general rule; since

---

[6] The overlap integrals form the 'inner products' of the linear space of the AOs ϕ_i. Due to a confusion between the two roles of differentials, the matrix S is sometimes called the 'metric' of the linear space. A metric m involving the ϕ_i must satisfy $m(\phi_1, \phi_3) \leq m(\phi_1, \phi_2) + m(\phi_2, \phi_3)$ and $m(\phi_1, \phi_2) = 0$ $(\phi_1 = \phi_2)$, hence $m(\phi_1, \phi_2) \geq 0$. Clearly the overlap matrix satisfies *none* of these requirements. A genuine metric can be defined in terms of S; the matrix $M = Z - S$ satisfies the above axioms where Z is a matrix containing unity in every position.

the optimum MO coefficients satisfy an effective one-electron eigenvalue equation (*11*), the equation

$$h^F T = ST\epsilon \qquad (27)$$

is technically as easy to solve as the orthogonal equation

$$\bar{h}^F \bar{T} = \bar{T}\epsilon \qquad (28)$$

(using the elementary transformation $\bar{h}^F = S^{-\frac{1}{2}} h^F S^{-\frac{1}{2}}$ and $\bar{T} = S^{-\frac{1}{2}} T$). Thus there is no pressing implementational reason for using an orthonormal basis in the MO model (UHF or RHF) and so this fact enables us to give separate consideration to the effects of numerical approximation in an orthogonal basis.

Orthogonalisation of the 'AO' basis has a rather dramatic effect on the feasibility of numerical approximation techniques. This is shown most clearly in the family of approximations known loosely as 'neglect of differential overlap' (NDO) techniques. The energy integrals arising in an orbital basis calculation (using the usual Born-Oppenheimer, spin-free Hamiltonian) are of two types: the kinetic energy integral

$$\int dv\, \phi_i - \tfrac{1}{2}\nabla^2\, \phi_j \qquad (29)$$

and the various 'electrostatic' integrals

$$\int dv\, \phi_i \tfrac{1}{r} \phi_j \; ; \quad \int dv_1 \int dv_2\, \phi_i(1)\, \phi_j(1) \tfrac{1}{r_{12}}\, \phi_k(2)\, \phi_l(2) \,. \qquad (30)$$

These latter integrals are of the form

$$\int dv\, V\, \phi_i \phi_j$$

where V is a multiplicative potential and involve the orbital products $\phi_i \phi_j$ which can be interpreted in the usual way as 'charge densities'. If we consider such an integral involving a non-orthogonal basis then, if ϕ_i and ϕ_j are remote from the source of the (Coulomb) potential V, the asymptotic form is

$$\int dv\, V\, \phi_i \phi_j \approx V \int \phi_i \phi_j = V\, S_{ij}$$

in particular, if $V = 1/r$ and the centroid of $\phi_i \phi_j$ is at distance R from the source of the potential, the asymptotic form is S_{ij}/R. Now if the basis is orthogonal the asymptotic form is zero unless $\bar{\phi}_i = \bar{\phi}_j$. Thus the orthogonal basis, since its charge densities $\bar{\phi}_i \bar{\phi}_j (i \neq j)$ contain no net charge, offers the possibility of a simple point charge model for many of the largest potential energy integrals and the approximation of *zero* for the majority of these integrals. The kinetic energy integrals are not covered by this approach but there are special approximation methods for these integrals involving an orthogonal basis (12). There remains the problem of the *evaluation* of the remaining orthogonal basis molecular integrals and this is discussed elsewhere (1). The fact that the 'off-diagonal' orbital products contain no net charge has an important interpretational advantage. The bonding integrals

$$\int dv \bar{\phi}_i \, \hat{h} \, \bar{\phi}_j \qquad\qquad (31)$$

are only dependent on the 'local' nuclei (at most) since the addition of an attracting centre remote from $\bar{\phi}_i, \bar{\phi}_j$ makes no contribution to the integral. Thus these integrals are much more 'environment independent' than the corresponding non-orthogonal equivalents where a contribution S_{ij}/R_x must be added for every additional nucleus X. From the point of view of systematic numerical approximation therefore, the strong orthogonality requirement is not a constraint but a considerable help in suggesting suitable schemes for investigation.

Turning to the effect of strong orthogonality on model approximations, we again use the H_2 example to make a point. It is well known that, like the single-determinant MO function, the complete CI expansion using a given basis is indifferent to linear transformations among the basis orbitals. So in our VB calculation on H_2 (reported in an earlier section) it is immaterial whether one used the AOs or an orthogonalised basis — the computed energy is invariant. However, it is equally familiar that there is no simple correspondence between the *terms* in the two expansions of the total wave function. In our case the largest contribution by far to the H_2 function is the homopolar Heitler-London function with spatial component $[\phi_1(1)\,\phi_2(2) + \phi_1(2)\,\phi_2(1)]$ if ϕ_1, ϕ_2 are the non-orthogonal 1 s functions. This function alone gives a total energy of -1.13636 a.u. compared to the total VB energy of -1.14518 a.u. In the case of the orthogonal basis the 'corresponding' function $(\bar{\phi}_1(1)\,\bar{\phi}_2(2) + \bar{\phi}_1(2)\,\bar{\phi}_2(1))$ alone gives a total energy of -0.61076 a.u. when orbital exponents are optimised and the optimum exponent is 0.927, completely different from the optimum AO value of 1.201. This result is obtained using the symmetrical orthogonalisation which gives an orthogonal basis as close as possible to the AO basis. It is perfectly clear that, at least for the electron-pair model, the strong orthogonality constraint has an enormous effect on the model approximation. Models which are chemically realistic in the AO basis become quite hopeless in the orthogonal basis. Of course, by using the 'ionic' structures the orthogonal basis recovers the energy but at the expense of a rather bogus model of the chemical bond as a 'charge transfer' phenomenon. This result turns out to be of quite general applicability for chemical bonding: if an orthogonal basis is used then a large number of chemically-unlikely-looking polar structures must be included in the calculation (13). Thus, it seems that, using an orthogonal basis, the wave function used must be sufficiently flexible to allow the effects of non-orthogonality to be re-introduced by the variation principle. This is a little too pessimistic since we can re-consider the strict strong orthogonality constraint (11) for a compromise. In the electron-pair model (at least) it is not technically difficult to perform the full VB calculation within each electron-pair and not attempt to allow for non-orthogonality between the pairs where we anticipate little 'binding'.

In spite of the numerous advantages of interpretation and calculation of molecular integrals, we must conclude that, at bottom, the orthogonal basis is a technical convenience. Since the orthogonal basis can only be used in a general model of molecular electronic structure if the details of that model contain a way of 'back transformation' to the non-orthogonal basis. In the MO model the transformation between the two bases is straight forward. In the VB model enough structures must be included to enable the spurious structures included by orthogonalisation be cancelled out by the variational process.

Choice of Orthogonal Basis: Hybridisation

In the last section we have constantly referred to an orthogonalised basis and yet made no specific proposals for the generation of such a basis. We now take up this matter. If a matrix V transforms the overlap matrix S to unity in the equation

$$V\dagger SV = 1 \tag{32}$$

then the basis

$$\bar{\phi} = \phi V \tag{33}$$

is orthogonal. Thus V is constrained by

$$V\dagger V = W\dagger = S^{-1}$$

and the solutions, V, to this equation are only determined up to a unitary transformation since writing UV in place of V gives

$$V\dagger U\dagger UV = V\dagger V = S^{-1} \tag{34}$$

if $U\dagger U = 1$. Obviously we can use this fact to choose the best possible orthogonal basis for some purpose. That is we have the matrix U at our disposal to (say) optimise some model or numerical approximation scheme (or both).

We dispose of the simplest problems first. Any orbital-basis theory of molecular electronic structure which purports to be interpretable as a theory of valence is committed to the use of atom-centred functions (or, at least, functions which go over into atomic orbitals for some values of their parameters).[7] We therefore wish to stay as

---

[7] This consideration excludes the free-electron model for example, and the Floating Spherical Gaussian Orbital approach in its simplest form.

close as possible (consistent with orthogonality) to a set of atom-centred functions: a set of functions with sufficient flexibility to recover the usual (approximate) atomic orbitals when the molecule dissociates into separate atoms.

We begin by using only the *linear* degrees of freedom contained in U to optimise our model and numerical approximations. We shall see that a careful consideration of these linear transformations suggests a natural generalisation of the usual AO basis in a way which enables us to use some of the conclusions of earlier sections where we discussed molecular symmetry. For a given set of non-orthogonal orbitals it is well-known from the work of *Löwdin (14)* that the simplest solution to equation (34), $V = S^{-1/2}$ is also the one which is optimum in the sense of generating the closest orthogonal basis possible to the original set. We therefore concentrate on finding optimum non-orthogonal bases and then apply the Löwdin orthogonalisation technique.

Using a basis of approximate AOs then the only linear transformations available to us which do not destroy the atom-centred nature of the basis are linear combinations of the AOs centred on the same nucleus: hybridisation. Now the conventional sets of hybrid orbitals used in elementary valence theory go a considerable way towards our goal of an optimum basis σ-bonded molecules of conventional structure. These hybrids (sp^3; sp^2; sp; $d^2 sp^3$ etc.) concentrate the non-orthogonality of the basis into pairwise overlapping orbitals and are a natural vehicle for the implementation of the electron-pair model. At the time of writing, however, there is something of an historical barrier to be overcome in putting forward the use of hybrid orbitals in quantitative theories of valence. There are positive and negative aspects to this barrier. Firstly, the quantitative dominance of the single-determinant MO model − with its invariance against linear transformations of basis − has of course tended to push the problem of optimum basis into the background. Secondly, the quantitative failures of early, rather naive, applications of simplified valence bond theory has left a residue of prejudice against the VB method itself *and*, unfortunately, against the use of hybrid orbitals. Notwithstanding these historical factors it is now established scientifically (*15*) that a basis of atom-centred orbitals with directional properties which enable a pairwise-overlapping basis to be used have considerable conceptual, numerical and model advantages over the raw AO set. The NDO approximations are quantitatively more rigorous when used in conjunction with this basis; the charge and bond-order matrix is much more readily interpretable using hybrids and the more general (non-MO) methods of calculating molecular electronic structures such as the electron-pair model are scarcely feasible in an ordinary AO minimal basis.

The numerical advantages of the hybrid basis arise from the 'concentration' of the non-orthogonality of the basis into local pairwise overlaps. Thus, if two orbitals are a 'bonded pair' then they overlap strongly and this overlap means that the orthogonalisation process effects them noticeably. If, however, two orbitals are not a bonded pair then the overlap between them is often much smaller and the effects of orthogonalisation between this pair are often negligible. Since all methods of generation of molecular integrals are for the non-orthogonal basis, the hybrid atomic orbitals (HAOs) have many integrals which are transferable from non-orthogonal to orthogonal basis. As we shall see in a later section on numerical approximation, the hybrid basis can be made only 'pairwise orthogonal' for the purposes of molecular integral generation.

The use of the Löwdin orthogonalisation technique (or any other method of orthogonalisation) means inevitably that the final basis of orthogonalised hybrid atomic orbitals (OHAOs) does contain 'many-centre' orbitals in the sense that each OHAO is mainly its HAO parent but necessarily contains (minimal) contributions from overlapping HAOs.

We have purposely not attempted to use the full freedom of the unitary matrix $\mathbf{U}$ in (34) since we wish to maintain a valence interpretation and to make use, wherever possible, of the transferability of molecular integrals. By using Ruedenberg's generalisation of the Mulliken approximation, *Roby (16)* has shown that, in the limit of a complete set on each atomic centre, the so-called canonically orthogonalised orbitals[8] have some formally attractive properties from the point of view of NDO approximations. In particular, for such a basis the NDO approximations are exact for the complete set on each atom. But the canonically orthogonal basis is symmetry adapted – that is, each member of the basis is a linear combination of symmetry orbitals of the same symmetry type. This means that the canonically orthogonal orbitals are strongly de-localised; indeed for symmetrical molecules (H_2, π system of benzene) they *are* the molecular orbitals. Thus, this basis is not at all well-adapted to numerical approximation schemes, to transferability of molecular integrals and, a fortiori, to realistic models of molecular electronic structure.

We therefore conclude that, for a combination of model, numerical and conceptual reasons the OHAO basis is well-adapted to a theory of valence. The hybrid orbital basis (for simple molecules) has a distinctive symmetry property: it carries a permutation representation of the molecular symmetry group; the equivalent orbitals are always sent into each other, never into linear combinations of each other. This simple fact enables the hybrid orbital basis to be studied in a way which is physically more transparent than the conventional AO basis.

---

[8] The canonically orthogonal set is obtained by choosing $\mathbf{U}$ in (34) to be the matrix which diagonalises $\mathbf{S}$. Hence [14] $\chi = \phi \mathbf{U} \mathbf{D}^{-\frac{1}{2}}$ where $\mathbf{D}^{-\frac{1}{2}}$ is the diagonal matrix of eigenvalues of $\mathbf{S}$. This set is particularly useful in identifying and eliminating redundancies in the original orbital basis ϕ.

Symmetry Considerations and the Hybrid Atomic Orbitals

We have seen earlier that, except for very small internuclear separations, an orbital basis with near-optimum scale factors (orbital exponents) does not have to be *constrained* to have the molecular symmetry. This conclusion was based on the rather slender evidence of the H_2 example and so we propose to apply this idea to our hybrid orbital basis and to test the validity of the ideas when applied to more realistic situations. With the exception of a few highly symmetrical molecules (chiefly belonging to the cubic point group symmetries), the set of hybrid orbitals 'pointing along the bonds' generated by the appropriate linear transformation among the AOs has *higher symmetry than the molecular environment requires*. Thus, in ethane, C_2H_6, there is no symmetry reason why the sp^3 hybrid involved in the C—C bond should be equivalent to the sp^3 hybrid pointing along the C—H bond. Indeed there is every reason to suspect that the optimum scale of the orbital basis along these two directions will be different. The lone-pair sp^3 hybrid in ammonia will, presumably, be much less affected by molecule formation than the sp^3 hybrids in the N—H directions. Now in qualitative valence theory these considerations are taken up in the Gillespie-Nyholm rules for electron-pair repulsions but, paradoxically enough, are ignored in quantitative minimal basis calculations which are the alleged theoretical background to the qualitative rules. In short it is formally unnecessary and intuitively wrong to use an orbital basis which has a higher symmetry than the molecule. Whatever the outcome of allowing the variation principle to operate without constraints it seems unduly restrictive to insist on a basis with spuriously high (atomic) symmetry: this symmetry over-restriction can surely be removed without loss. Again, we must note here that this rather obvious statement has enormous repercussions on conventional semi-empirical techniques of calculation of molecular electronic structure. It means, for example, that atomic data can only rarely be used as a substitute for molecular integrals since the atom-in-molecule orbitals are not the same as the separate atom orbitals — worse, they are no longer equivalent among themselves. An 'atomic' self-repulsion integral $(\phi_i \phi_i, \phi_i \phi_i)$ is different if ϕ_i is the lone-pair hybrid of NH_3 or the bond-pair hybrid as the Gillespie-Nyholm rules suggest.

The most difficult problem we face in deciding to use a basis of hybrids which reflects the molecular symmetry is: how do we choose such a basis in view of the enormous numerical difficulties involved in optimising the non-linear parameters in molecular calculations? The real question is: are there any *rules* for this choice, can the optimisation be done (at least approximately) once and for all? The chemical evidence is for us — it is the most basic concept of the theory of valence that particular electronic sub-structures tend to be largely environment-independent. How can we select our basis to reflect this chemical fact?

Generalised Hybrid Orbitals

The use of hybrid atomic orbitals in qualitative valence theory has, in the past, rested on two points: (i) an empirical 'justification' of their use involving the concept of the 'valence state' of an atom and (ii) a simple linear transformation technique for the construction of the explicit forms of the orbitals. In this section we show that both of these points can be replaced. The justification can be replaced by a derivation and this derivation can be used to suggest variational forms which render the linear transformation technique redundant.

In the classical construction, a molecular geometry is regarded as 'given' and a linear transformation is sought which generates a set of hybrid orbitals along or close to the bond directions from the 'raw' atomic orbitals: i.e. the complex or (more usually) the real atomic orbitals which are eigenfunctions of $\hat{H}$ and $\hat{L}^2$. Thus, a given hybrid (h_i) is written as

$$h_i = \sum_j X_{ji}\, \phi_j \qquad (35)$$

where X is a unitary (orthogonal) matrix of hybridisation coefficients and the ϕ_j are the AOs. We can proceed in either of two ways to generalise this definition. We can study the forms of the AOs ϕ_j and their contribution to h_i (15) or we can re-examine the reasons for using a basis of AOs in the first place (17).

The first method involves us in a specific choice for the forms of the ϕ_j, let us choose STOs since there is ample evidence for their suitability as molecular basis functions. Then, if we choose our co-ordinate axes in a suitable way, each hybrid based on a first-row atom (Li $-$ F) can be written

$$h_i = \sin\alpha(2\,s) + \cos\alpha(2\,p_\sigma) \qquad (36)$$

(for some value of the hybridisation parameter α).

Orbitals of this form are said to be sp^x hybrids (where $x = \cot^2\alpha$). If now the (2 s) and (2 p) functions are STOs we have

$$(2\,s) = \sqrt{\frac{1}{2\,\pi}\frac{(2\,\zeta)^{5/2}}{\sqrt{2}}}\; r\,\exp(-\zeta r)$$

$$(37)$$

$$(2\,p_\sigma) = \sqrt{\frac{1}{2\,\pi}\frac{3}{\sqrt{2}}\frac{(2\,\zeta)^{5/2}}{\sqrt{2}}}\; r\cos\theta\,\exp(-\zeta r)$$

and so, provided that the orbitals (2 s) and (2 p) have the same orbital exponent (ζ):

$$h_i = N\left(\frac{\sin\alpha}{\sqrt{3}} + \cos\alpha\,\cos\theta\right) r\,\exp(-\zeta r)$$

or, for the corresponding (ns), (n p_σ) orbitals of any shell:

$$h_i = N\left(\frac{\sin\alpha}{\sqrt{3}} + \cos\alpha\,\cos\theta\right) r^{n-1} \exp\left(-\zeta r\right). \tag{38}$$

This function can be written

$$h_i = f_i(z, r) \exp\left(-\zeta r\right) \tag{39}$$

since $z = r\cos\theta$, is clearly cylindrically symmetrical about the 'bond' axis and its scale is determined by the exponent. The functions $f_i(z, r)$ in (39) determine the 'type' of hybridisation (sp, sp^2, sp^3 or intermediate types), leaving $\exp\left(-\zeta r\right)$ to determine the 'size' of the hybrid. Clearly, it is a restriction on the atomic wave function to constrain the (ns) and (np) STOs to have the same exponent but in a molecular situation the form (38) or (39) suggests a generalisation. If we choose a *separate* orbital exponent for each hybrid we immediately obtain up to four non-linear parameters with which to optimise the minimal basis which should more than compensate for the simplification of (37) to obtain (38). More than this, the use of a separate scale factor for each hybrid enables us to use a set of atomic orbitals which have the true symmetry of the molecular environment (or lower if we wish to pursue the 'symmetry dilemma in a more general context than H_2). Thus we shall refer to the basis (39) as a basis of Generalised Hybrid Orbitals (GHOs) and, for the time being fix the *forms* of the $f_i(z, r)$ by the traditional rules for hybrids-along-bonds; we can bear in mind that (e.g.) the hybridisation parameter α in (38) is available to be optimised if we choose.

It is immediately obvious that, in addition to their variational advantages, the GHOs have a number of convenient conceptual advantages over the conventional 'spherical' hybrid orbitals.

(i) The form of the functions and the values of the non-linear parameters passes over smoothly into the separate-atom AOs as a given atom is taken from a molecular environment.

(ii) The destruction of spherical symmetry occurring as an atom is introduced into a molecular environment is reflected in the form of the functions and the differences among the GHO exponents. This contrasts strongly with the usual spherically-based AOs where the removal of spherical symmetry cannot be allowed for in the minimal basis; only by using an extended 'AO' basis can these effects be introduced and then only in a way which is extremely difficult to interpret. The construction of an orbital along a particular direction is measured directly in the GHO case by an increase in the corresponding ζ_i. When using an extended basis this phenomenon has to be inferred by examination of a mass of *linear* coefficients

(iii) The very idea of 'local atomic' re-arrangements of charge occurring as molecules are formed (*18*) is outside the scope of most minimal basis calculations – the whole burden of the variational calculation falls on the linear coefficients in the

MO or VB models. Occasionally atom-in-molecule orbital exponents are used (principally for H atoms, using 1.2) but it is unusual to see any interpretation of this fact.

The upshot of these points is that, in throwing the emphasis on extending the 'AO' basis and thereby generating a mass of linear coefficients by the variation process, much of the interpretation of the phenomena occurring on bond formation is rendered very difficult. It is not easy, for example, to unambiguously separate the three effects discussed above from each other.

The second approach to the definition and use of GHOs is less familiar but ultimately more satisfactory. It depends on the generation of the hybrid orbitals as solutions of the hydrogen-atom Schrödinger equation which are not eigenfunctions of the 'usual' commuting operators $\hat{L}^2$ and $\hat{L}_z$. It has been shown elsewhere (17) that the hydrogen-atom Schrödinger equation is separable in prolate spheroidal coordinates with the nucleus at one focus of the ellipsoids of revolution. In this system the commuting set of operators are $\hat{H}, \hat{L}_z$ and

$$\hat{A} = \hat{L}^2 + \frac{R^2}{4}(\hat{p}_x^2 + \hat{p}_y^2) - \frac{Rz}{r_1}$$

where $\hat{L}^2$ is the total angular momentum operator, $\hat{p}_x$ and $\hat{p}_y$ are the Cartesian linear momenta and r_1 is the distance from the nucleus.

The eigenvalues of $\hat{H}$ and $\hat{L}_z$ are, of course, the familiar $-1/2\,n^2$ and m respectively while the simultaneous eigenfunctions and the eigenvalues of A depend on R (the distance from the nucleus to the other focus of the co-ordinate system). This last quantity is arbitrary and we can generate whole families of eigenfunctions of the hydrogen-atom Hamiltonian by varying the parameter R. In particular, the limits R = 0 and R = ∞ generate respectively the familiar spherical polar co-ordinates and the parabolic co-ordinates. Both of these limiting cases are known to separate the hydrogen-atom Schrödinger equation.

Intermediate values of R have eigenfunctions which are linear combinations of the usual (complex) hydrogenic orbitals with the same values of n and m: hybrid atomic orbitals. The actual values of the linear combination coefficients determining the explicit form of the hybrids depend on R: in the case of n = 2 m = 0 we get an inequivalent pair of hybrids pointing in opposite directions of the general form

$$h_1 = \sin \alpha(2\,s) + \cos \alpha(2\,p)$$
$$h_2 = \cos \alpha(2\,s) - \sin \alpha(2\,p)$$

where α is determined by R. The detailed derivations are given in (17). For our purposes three properties of these hydrogenic hybrid orbitals are important:

(i) Their very existence: hybrid orbitals are generated naturally by the solution of the hydrogen-atom Schrödinger equation. They are on an *equal theoretical footing* to the more familiar complex atomic orbitals.

(ii) Their dependence on R: if we take the obvious choice for the interfocus distance — the bond length of a particular model system — then this bond length alone determines *all* the σ, π, δ etc. hybrids for each atom. It may well turn out, particularly for many electron systems, that these hybrids are not the optimum ones but the possibility of a unique set is available.

(iii) Their properties as zeroth-order functions in perturbation theory: it is not surprising that these orbitals are the correct zeroth-order functions for the perturbation of a hydrogen-atom by a point charge — they are the obvious candidates to use in crystal- and ligand-field theories of molecular structure.

To return to our main line of thought, the development of variational techniques for valence theory, there is an obvious parallel between the use of the usual (complex) AOs for the calculation of atomic electronic structures and the hybrid AOs for the molecular case:

Commuting Operators	$\hat{H}, \hat{L}^2, L_z$	$\hat{H}, \hat{L}_z, \hat{A}$
eigenfunctions	hydrogenic AOs	hydrogen hybrids
Variational 'Generalisation'	STOs	GHOs
Electronic Structure	spherically symmetry atoms	Molecules with local cylindrical symmetry ('bonds')

We have therefore established the theoretical pedigree of GHOs and so propose to investigate their use in molecular electronic structure calculations.

First, however, we must note some properties of the GHOs which differ from the ordinary hybrid orbitals. Most important, they cannot (except in the case of highly symmetrical molecules) be generated from a single set of (e.g. ns, np nd) AOs. Thus, in general, they are not orthogonal — each GHO will have a small non-zero overlap integral with other GHOs *on the same centre*. Also, a point to which we will return later, they can be optimised separately and so the choice of an optimum GHO basis is a substantive issue even in the single-determinant MO method: e.g. the optimum 'bent bond' hybrids of C_2H_4 will not be related to the optimum $\sigma - \pi$ hybrids by a linear transformation.

It is now clear why we had to establish the superiority of the UHF method over the RHF method. There is little point in using a basis of GHOs whose separate-atom asymptotic form is the atomic orbitals if they are used in the RHF method which specifically prevents the separate-atom optimum atomic orbitals from being recovered. The advantages of the GHOs (or *any* molecule-optimised orbitals) may fall to the ground if they are used in the RHF method (except where the RHF results coincide with the UHF results around equilibrium). The spin-eigenfunction constraint potentially neutralises any effort to use orbitals containing a non-linear variational parameter since, at inter-nuclear separations away from equilibrium, the spurious limit created by this spin-eigenfunction constraint distorts the form and scale of the separate atom orbitals (cf. H_2 above; optimum RHF exponent at $R \rightarrow \infty$ of 27/32). In everything that follows in this section we tacitly use a *realistic* model for the molecular electronic structure at all internuclear separations (e.g. UHF or electron-pair). Deferring questions of numerical techniques for the moment to a later section, we can investigate these functions as basis functions for the calculation of molecular electronic structure *and* investigate the spatial symmetry constraint in a wider context than the hydrogen molecule.

In order to limit the amount of computer time we have concentrated on a few simple molecules: the first-row hydrides CH_4, NH_3, H_2O and two molecules chosen to exhibit a range of bond types: CH_2O and BH_3CO. In the case of the hydrides the calculations were performed to investigate the effect on orbital exponents of a 'homologous' series and to obtain more information about the spatial 'symmetry dilemma'. All calculations were performed in the first instance at the experimental nuclear geometry using the UHF method. In all three cases (CH_4, NH_3, H_2O) a basis of four sp^3-type orbitals was used on the central atom in addition to a 1 s core function. For CH_4 these functions already point along the bonds, in NH_3 and H_2O the 'bonding' hybrids pointed along the bonds. For H_2O, where a choice is necessary, the lone pair hybrids were chosen to point along the conventional lone pair directions.

All the basis functions were given independently variable orbital exponents (CH_4, 9; NH_3, 8; H_2O, 7) and all exponents were optimised by the quadratically convergent direct search method of *Fletcher* (*19*). For comparison, the calculations were repeated with the GHOs constrained to have the symmetry of the molecule: three independent variables for CH_4 (1 s_C, sp^3, 1 s_H); four for NH_3 (1 s_N, sp^3, sp^3, 1 s_H) and four for H_2O (1 s_O, sp^3, sp^3, 1 s_H). The most striking qualitative result is the confirmation of the results quoted earlier for H_2: when the orbital exponents are all optimised, the GHO basis has the symmetry of the molecule: there is no spatial symmetry dilemma.[9]

If we now consider the numerical results quoted in Table 1 for the optimum exponents, three conclusions follow immediately. Firstly, the 1 s orbital on the heavy atom is unchanged by molecule formation; this is to be expected. Second, the sp^3 orbitals involved in the X–H bond are all *contracted* with respect to their free-atom values. Finally, the sp^3 orbitals containing 'lone pairs' of electrons are largely unchanged or expanded slightly on molecule formation. In fact, of course, the optimum separate atoms minimal basis functions do not have the same orbital exponent for the 2 s and 2 p AOs. To facilitate comparisons therefore in Table 1 the optimum n = 2 exponent is given for the atoms when such a constraint is imposed (the qualitative conclusions are, in any event, unchanged by use of these exponents for comparison or a notional exponent of 1/4 ($\zeta_s + 3\zeta_p$) or any reasonable choice).

---

[9] Strictly speaking, there is one minor qualification to be made to this general conclusion. In the case of CH_4, nine independent variables render the optimisation process numerically rather redundant – the strict conclusion is therefore slightly weaker; for no GHO basis with a symmetry lower than that of the molecule was an energy found lower than that of a symmetrical basis.

Table 1

XH_n	Orbital Exponents				
	$1 s_X$	sp^3 (bond)	sp^3 (lone pair)	$1 s_H$	atom X^a)
CH_4	5.6735	2.0245	–	1.0354	1.5693
NH_3	6.6687	2.3228	1.7873	1.1082	1.9090
H_2O	7.6673	2.6266	2.2028	1.2007	2.2318

a) Optimised orbital exponents for n = 2 shell of atom X constraining $\zeta_{2s} = \zeta_{2p}$: this involves an energy penalty of ~ 0.1 a.u.

It is also evident from Table 1 that the exponents of the hydrogen 1 s orbitals are increased (showing an orbital contraction effect) by different amounts in the sequence CH_4, NH_3, H_2O. It is interesting to attempt to establish some trends in these contractions even with only a small amount of data to hand. It is clear that, for the hydrogen atom orbital, the electronegativity of the atom X (or atomic number in this simple series) is important – the greater the electronegativity of X, the greater the orbital contraction. In the case of the 'bond' sp^3 orbital the orbital contraction – as measured by the difference between the molecular and atomic exponents or by the ratio – is roughly constant but there is a definite tendency for the contraction to be smaller for the heavier atom. Obviously the magnitude of the contraction effect is dependent on (at least!) the hybrid type (sp, sp^2, sp^3), the electronegativity difference between the two bound atoms (or orbitals) and the internuclear separation.

Clearly, if the GHOs are to be used in a theory of valence we must establish some rules for the approximate contraction (or expansion) factor which will enable us to avoid the tedious business of non-linear optimisation processes. This is work for the future – for the moment we simply investigate some of the properties of the GHOs when used to investigate the electronic structure of molecules at equilibrium geometry.

The molecules CH_2O and BH_3CO show a wider variety of electronic sub-structures than the simple hydrids discussed above: in particular CH_2O contains a (polar) π-bond and sp^2 hybrids and BH_3CO has π bonds, sp hybrids and a dative bond. It is of some interest therefore to see how the GHOs behave in these situations. Table 2 contains the relevant orbital exponents. The pattern of orbital contraction for GHOs involved in σ X–H bonds is repeated and reinforced by the C–O σ bond orbitals.

Table 2. Optimised GHO exponents for CH_2O and BH_3CO

Molecule	GHO exponents and bond type^a)										
	σ_{CH}	σ_{HC}	σ_{BH}	σ_{HB}	σ_{CO}	σ_{OC}	π_{CO}	π_{OC}	σ_O	σ_{BC}	σ_{CB}
CH_2O	1.8660 (sp²)	1.0713	–	–	1.7301 (sp²)	3.7617 (sp²)	1.4295	2.0632	2.2372 (sp²)	–	–
	–	–	1.5703 (sp³)	1.0062	1.7227 (sp²)	2.5206 (sp²)	1.5319	2.2160	2.2620 (sp)	1.2733 (sp³)	1.5659 (sp)

a) The notation is: σ_{XY} is the σ-type GHO centred on X 'pointing at' Y; π_{XY} is the π-type GHO (AO) on X bonded to Y; σ_X is a σ-type lone pair on atom X.
The 1 s shells are omitted.

Again the lone pair orbitals are substantially unchanged on molecule formation. In both molecules it is notable that the 2 p AOs involved in the π bonds are expanded with respect to the free atoms. Also noteworthy is the difference in behaviour between the GHOs involved in classical covalent σ bonds and those in the B $\leftarrow$ C dative bond. There is no appreciable orbital contraction of the sp GHO on carbon on forming the dative link. Although we cannot hope to make valid generalisations on the basis of the small amount of data reported here, it is most encouraging to see patterns of orbital contraction and expansion emerging which show qualitative differences among the chemically different electron pairs. It is also encouraging to note that, where comparison with existing methods is possible, our results are in qualitative agreement with ordinary AO methods: the difference between the diffuse $2 p_\pi$ AO and the more compact $2 p_\sigma$ AOs in CH_2O has been noted.

In order to investigate the possibility of obtaining 'empirical' rules for the contraction or expansion of the GHOs in various molecular environments and for various internuclear separations, a good deal of non-empirical calculation will have to be done. It is necessary to discover the essential variables governing the optimum scale factors for GHOs. For the present we simply note the qualitative conclusions above and the general agreement between our results and Ruedenbergs detailed analysis of the binding in H_2^+, (18). We would hope in the future to be able to present an analysis of the energetics of individual bond formation based on the behaviour of the GHOs.

In summary, the GHO basis (with optimum scale factor for each orbital), together with a suitable variational model (UHF, electron-pair) provides an extremely useful minimal basis for a theory of valence. The concepts used to interpret the various phenomena occurring on bond formation can be given some quantitative meaning and the various 'constraints' on the wave function can either be abandoned as physically irrelevant or are satisfied automatically by the GHO basis. No doubt with the excessively restrictive spin-eigenfunction constraint in mind, Primas (20) has called the UHF method 'robust' in the sense of being a stable representation of the physics over a wider range of conditions. Borrowing his terminology we might therefore contrast the 'robust' nature of the GHO basis with the 'non-robust' nature of the ordinary (exponent-fixed, spherically-based) AO basis.

We have made one rather obvious omission from our descriptions of molecule electronic structure – the structure of transition-metal ions. This is deliberate since, in spite of the well-developed theories of the electronic *spectra* (U.V., photo-electron) of these compounds, it is still true to say that there is no theory of the *bonding* in this important class of molecules. The question of the localised or de-localised nature of the electronic structure of the bonds in these systems has not really been solved: historically, there has been some skirmishing about the superiority of the MO or VB *methods* but the nature of the valence in these molecules has received a disproportionately small amount of attention. Thus any attempt to develop a GHO basis for transition-metal compounds is perhaps premature until more experience has been gained with typical element chemistry.

Numerical Methods for the GHOs

The GHOs are related to the STOs and so the standard techniques of molecular integral evaluation for STOs (*21*) can be used for the GHO molecular integrals. Many of the existing implementations of the STO integral formulae, however, are not suitable since, in order to conserve resources, these programs do not have sufficient flexibility to allow linear combinations of STOs of different l value or different orbital exponents for different STOs of a given l value (*22*). Thus we have developed simple Gaussian-expansion methods for the GHO integrals. The qualitative form of hybrid orbitals is a series of 'lobes' (of either positive or negative sign) whose centres lie along a line — the 'direction' of the hybrid. This functional form can be approximated by using a single Gaussian function ('s-type', form $\exp(-\alpha r^2)$) at the centroid of each lobe and optimising the Gaussian exponent for each lobe together with the linear combination coefficients for the approximation of the GHO by these lobe functions. The details of the technique are given elsewhere (*15*) — briefly the utility of this technique hinges on two 'scaling relations'. Firstly, the familiar scaling relation for a Gaussian fit to an STO function: the exponents and linear combination coefficients are essentially independent of the STO exponent (*23*) so that one fitting procedure is needed for each type of STO. Secondly, the positions of the centroids of the lobes of a GHO are determined by the GHO orbital exponent and it turns out that a scaling relation exists for these centroids. Thus a given type of GHO (form of $f(z, r)$ in (39)) can be fitted by a linear combination of spherical Gaussians *once* and *for all*.

It must be stressed that the use of GHOs in no way depends on this latter Gaussian expansion procedure since all molecular integrals reduce to standard STO forms. It has merely proved expedient to make use of the expansion method in calculations reported earlier. This ab initio technique provides the raw data with which to establish the patterns of behaviour of the GHOs. We can now address ourselves to one of our stated aims: the development of approximation methods for a quantitative theory of valence.

The GHO Basis, Approximation Methods and the NDO Approximation

The "Neglect of Differential Overlap" (NDO) family of approximations is based on the orthogonalised atomic orbital basis: in order to be interpretable the NDO approximations must be made in an AO basis for which the overlap densities (differential overlaps) contain no net charge, i.e. integrate to zero. Now in using a basis of 'ordinary' hybrids it has been shown that this approximation family performs best in a *localised orthogonalised* hybrid basis where the non-orthogonalities among the AOs is concentrated into well-defined regions. This concentration has the effect of increasing the overlap integral between 'bonded pairs' of orbitals and so making these orbitals change by quite large amounts during the orthogonalisation procedure. Typical $\sigma-\sigma$ overlaps between pairs of bonded hybrids are $0.6 - 0.8$. Some idea of the change induced by orthogonalisation can be had by considering a 'model' of two overlapping AOs ϕ_1, ϕ_2 with overlap integral 0.7: the corresponding orthogonalised orbitals are:

$$\bar{\phi}_1 \cong 1.30\,\phi_1 - 0.53\,\phi_2$$
$$\bar{\phi}_2 \cong 1.30\,\phi_2 - 0.53\,\phi_1$$

these are the orthogonal AOs closest possible to the originals. Thus $\bar{\phi}_1$ and $\bar{\phi}_2$ are quite different from ϕ_1 and ϕ_2 and the molecular integrals involving the AOs ϕ_1 ϕ_2 suffer quite large changes on orthogonalisation (1). This means that any method of approximation for the molecular integrals which makes use of non-orthogonal AO integrals plus 'corrections' is likely to be fairly difficult to apply. The 'corrections' to the two-centre electron repulsion integrals, for example, can be up to 15%.

When the GHOs are used as a basis we obtain an important 'bonus' due to the orbital contraction effects. In the critical σ bond case the systematic contraction of GHOs means that the overlap integral between a bonded pair ϕ_1, ϕ_2 is much smaller than the corresponding integral for the conventional hybrids: the $sp^3 - 1\,s_H$ overlap falls from 0.787 to 0.612 for CH_4; 0.683 to 0.563 for NH_3; 0.568 to 0.503 for H_2O; and similar reductions for other σ-bonded pairs. Since the overlap integral falls, the orthogonalisation correction falls and so therefore does the dependence of molecular integrals on the orthogonalisation process. *We can therefore see that the GHOs are well-adapted to the NDO approximation scheme: they have all the advantages of the conventional hybrid basis plus an important lowering of dependence on orthogonalisation.* Thus all the results which show that an orthogonalised hybrid basis is an optimum basis for the use of the NDO family of approximations apply to the orthogonalised GHO basis. We now go on to suggest some approximation schemes which make use of the properties of the GHO basis. These approximation schemes are provisional and are *not* part of our general chemical and methodological arguments in favour of the GHO basis. The principal justification for the GHO basis is its validation of the qualitative concepts of valence theory.

It is a good approximation to treat the GHOs as an orthogonal set except for the bonded pairs: the non-bonded GHO overlaps are all less than 0.1 for the valence orbitals of CH_4, NH_3 and H_2O. We can then study the effect of the orthogonalisation process for the *bonded pairs alone* by explicitly performing the 2×2 orbital transformation (12). If the two GHOs of a bonded pair generate a local 2×2 overlapping matrix

$$S = \begin{bmatrix} 1 & S \\ S & 1 \end{bmatrix}$$

then the corresponding Löwdin orthogonalisation matrix is

$$V = \begin{bmatrix} V_{11} & V_{12} \\ V_{21} & V_{22} \end{bmatrix}$$

where

$$V_{11} = V_{22} = \frac{1}{2\sqrt{1-S^2}}\,(\sqrt{1-S}+\sqrt{1+S})$$

and

$$V_{12} = V_{21} = \frac{1}{2\sqrt{1-S^2}}\,(\sqrt{1-S}-\sqrt{1+S})$$

Thus the differential overlap factor $\bar{\phi}_1\,\bar{\phi}_2$ for the orthogonalised GHOs is

$$\bar{\phi}_1\,\bar{\phi}_2 = \frac{1}{(1-S^2)}[\phi_1\,\phi_2 - \frac{S}{2}(\phi_1^2 + \phi_2^2)]$$

and one cannot avoid the temptation to use the Mulliken approximation for the non-orthogonal overlap density

$$\phi_1\,\phi_2 = \frac{S}{2}(\phi_1^2 + \phi_2^2)$$

and so

$$\bar{\phi}_1\,\bar{\phi}_2 \equiv 0\,.$$

The pairwise overlap, symmetrical orthogonalisation and Mulliken approximation together validate the NDO approximation — the orbital product in the orthogonalised GHO basis vanishes to the extent that the Mulliken approximation is realistic. This conclusion obviously has enormous consequences for any NDO approximation schemes.

The off-diagonal one-electron integral ('core' integrals)

$$\int dv\,\bar{\phi}_1\,\hat{h}\bar{\phi}_2 = \bar{H}_{12}$$

are combinations of kinetic and potential energy terms since

$$\hat{h} = -\frac{1}{2}\nabla^2 + \sum_{\alpha} \hat{V}_{\alpha}$$

where $\hat{V}_{\alpha}$ is the core potential of centre α. Now apart from very small core-valence exchange terms, each $\hat{V}_{\alpha}$ is simply multiplicative so that, using $\bar{\phi}_1\,\bar{\phi}_2 = 0$ we have

$$\int dV \, \bar{\phi}_1 \, \hat{V}_\alpha \bar{\phi}_2 = 0 \qquad (\alpha = 1, 2;) .$$

Thus, the only remaining term in the integral is the kinetic energy term

$$\int dV \, \bar{\phi}_1 (-\tfrac{1}{2}\nabla^2) \, \bar{\phi}_2 .$$

Calculations show that this kinetic energy integral is, for example, negative for a bonded pair, $\bar{\phi}_1, \bar{\phi}_2$ and does approximate the value of $\bar{H}_{12}$ rather well (12)..

A similar analysis of the diagonal one-electron integrals shows that to a first approximation, the potential energy contribution to $\bar{H}_{11}, \bar{H}_{22}$ is *unchanged* by symmetrical orthogonalisation. Only the kinetic energy term suffers appreciable change in the transition from the GHO basis to the corresponding orthogonalised basis (12)

$$\bar{H}_{11} \quad \bar{T}_{11} + V_{11}$$

where

$$\bar{T}_{11} = \int dV \, \bar{\phi}_1 (-\tfrac{1}{2}\nabla^2) \, \bar{\phi}_1$$

and

$$V_{11} = \int dV \, \phi_1 (\sum_\alpha V_\alpha) \, \phi_1$$

The largest electron repulsion integrals

$$\gamma_{ii} = \int dv_1 \int dv_2 \, \phi_i^2(1) \, \frac{1}{r_{12}} \phi_i^2(2)$$

$$\gamma_{12} = \int dv_1 \int dv_2 \, \phi_1^2(1) \, \frac{1}{r_{12}} \phi_2^2(2) .$$

also respond to an alalysis using the Mulliken approximation. The transformation to the orthogonal basis and the use of the Mulliken approximation gives

$$\bar{\gamma}_{11} = A \, \gamma_{AV} - B \, \gamma_{12} + C \, \Delta\gamma$$
$$\bar{\gamma}_{22} = A \, \gamma_{AV} - B \, \gamma_{12} - C \, \Delta\gamma$$
$$\bar{\gamma}_{12} = A \, \gamma_{12} - B \, \gamma_{AV}$$

where

$$A = \frac{1}{2}\left(1 + \frac{1}{1-S^2}\right); \quad B = \frac{S^2}{2(1-S^2)}; \quad C = \frac{2}{1-S^2}$$

$$\gamma_{AV} = \frac{1}{2}(\gamma_{11} + \gamma_{22}); \quad \Delta\gamma = \frac{1}{2}(\gamma_{11} - \gamma_{22}) .$$

It is shown in (*12*) that, by using the approximation $\gamma_{11} \cong \gamma_{22}$ and assuming that, in the region of large overlap, $\gamma_{12} \cong (1 + S^2)\, \gamma_{AV}/2$ we obtain

$$\bar{\gamma}_{ii} = \frac{1}{1 - S^2}\, [1 - \frac{S^2}{4}(3 + S^2)]\, \gamma_{ii}$$

$$\bar{\gamma}_{12} = \frac{1}{1 - S^4}\, [1 - \frac{S^2}{2}(1 + S^2)]\, \gamma_{12}\, .$$

This analysis therefore provides some theoretical justification for the rather empirical 'scaling rules' which have been used to simulate the effect of orthogonalisation in earlier work (*5, 24*).

In summary, the GHO basis is one in which the NDO approximation is most accurate: the localisation and reduction of the overlap integrals minimises the effect of the orthogonalisation procedure. However, as we pointed out earlier, none of the empirical techniques for generating molecular integrals from experimental data can be applied: the GHOs are atom-in-molecule orbitals. Thus any approximation techniques for the generation of the retained integrals in the NDO scheme must be non-empirical. Using the GHO basis therefore forces us to abandon the empirical techniques in favour of the approximation of integrals over an *explicitly-defined orbital basis.*

Molecular Shapes, the VSERP Model, and the GHO Basis

We have so far used the GHO basis as the ideal atomic orbital basis for a valence theory of molecular electronic structure. These hybrid orbitals are a natural variational extension of the solutions of the hydrogen atom Schrödinger equation and, as such, are not confined to being generated from the familiar (complex) atomic orbitals by a unitary transformation. So, for example, there is no reason why these should be four sp^3-type GHOs centred on any atom nor that any such sp^3-type GHOs should make angles of $109° \, 28'$. The derivation we have given depends not on the mathematical properties of the rows and columns of a unitary 'hybridisation matrix' but on the local (two-centre) potential field experienced by the electrons in a bond: that is, on the physics of the situation. We are therefore, when using a GHO basis, completely free of the usual restriction imposed on a hybrid basis: the nature of the GHOs (sp^x, x real) *need not fix the mutual directions* of the GHOs. We are concerned with the GHOs principally as an optimum minimal basis for the electron-pair bond, that is, the local diatomic (or monatomic for lone pairs) potential is the important factor. Thus, whatever model we choose for the details of the calculation (MO, VB) of the electronic structure of the electron pairs in a molecule, the problem of the mutual orientations of the electron-pairs is still to be determined. Having developed an essentially localised electron pair model of molecules it is clear that the interactions between these pairs determine the overall shape of the molecule. This problem is then treated by the variation principle — we simply find a suitable arrangement of the electron pairs which minimises the total electronic energy.

When an atom has two (or more) lone pairs of valence electrons the problem of the justification of the usual chemical ideas is more acute than with bond pairs. Apart from the familiar chemical evidence, there is no theoretical reason to assume that (e.g.) the four electrons of H_2O not involved in the localised O—H bonds do indeed take up 'paired' distributions. When the computed wave function for H_2O are analysed the distribution of electrons in the non-bonding region is smooth, showing no maxima in the conventional near-tetrahedral directions. Many workers have used the invariance of the single-determinant MO wave function to generate localised lone-pair orbitals and so attempt to justify the chemical ideas. But what is needed is precisely the opposite: a demonstration of the existence of lone pairs of electrons when the transformation to a localised description is *not* arbitrary. That is, when there is some (preferably energetic) *criterion* for the existence of lone pairs. It is the very arbitrariness of the localised MO description which undermines its value. The principal piece of chemical evidence for the existence of lone pairs is the readiness with which many molecules containing these assumed structures form electronpair dative bonds. But this readiness to form such bonds does not necessarily imply the existence of lone pairs *as such* in the donor system; it merely shows that pairs of electrons are easily localised by a suitable acceptor. In spite of these latter considerations we will assume that the existence of lone pairs *is* established and investigate their interactions.

The most widely used qualitative model for the explanation of the shapes of molecules is the Valence Shell Electron Pair Repulsion (VSEPR) model of *Gillespie* and *Nyholm (25)*. The orbital correlation diagrams of Walsh *(26)* are also used for simple systems for which the qualitative form of the MOs may be deduced from symmetry considerations. Attempts have been made to prove that these two approaches are equivalent *(27)*. But this is impossible since Walsh's Rules refer explicitly to (and only have meaning within) the MO model while the VSEPR method does not refer to (is not confined by) any explicitly-stated model of molecular electronic structure. Thus, any 'proof' that the two approaches are equivalent can only prove, at best, that the two are equivalent *at the MO level* i.e. that Walsh's Rules *are contained* in the VSEPR model. Of course, the transformation to localised orbitals of an MO determinant provides a convenient picture of VSEPR rules but the VSEPR method itself depends not on the independent-particle model but on the possibility of separating the total electronic structure of a molecule into more or less autonomous electron pairs which interact as separate entities *(28)*. The localised MO description is merely the simplest such separation: the general case is our Eq. (6)

$$\Phi = \hat{A}_x \prod \Phi_R$$

where each Φ_R is a pair-function. We therefore unblushingly commandeer the VSEPR model as the natural extension of our theory of valence based on the pairwise bonding of GHOs.

Before investigating the qualitative concepts of the VSEPR model it is worth noting that the details of the interactions between the electron pairs have been ascribed to a 'size-Pauli exclusion principle result'. But objects do not repel each other simply because of their sizes (i.e. interpenetrations); only if the constituents of the objects interact is any interaction possible[10]). If we are to use the idea of orbital size at all we must avoid the danger of *contrasting* a phenomenon (electron repulsion) with one of its manifestations (steric effects). The only quantitative tests which we can apply to the VSEPR model are ones based on the terms in the molecular Hamiltonian; specifically, electron repulsion.

The key to the utility of the VSEPR model is, of course, the repulsion sequence

$$\text{lone pair} - \text{lone pair} > \text{lone pair} - \text{bond pair} > \text{bond pair} - \text{bond pair}$$

how does this hold up in practice? The water molecule is the simplest system which enables the whole sequence to be evaluated. Using sp^3-type GHOs with separate orbital exponents for the distinct hybrids on oxygen (bond, lone pair) we obtain, for the optimised GHOs,

lone pair – lone pair	2.452	a.u.
lone pair – bond pair	2.186	a.u.
lone pair – bond pair	1.963	a.u.

for the inter-pair repulsion energies in the electron pair model (VB in each bond).

The pair-pair repulsion energies E^{AB} are defined by

$$E^{AB} = \text{trace } \mathbf{P}^A \, \mathbf{G}^{AB}$$

where $\mathbf{P}^A$ is the 'charge and bond-order matrix of pair A and

$$(G^{AB})_{ij} = \sum_{tu} P^B_{tu}[(ij, tu) - \tfrac{1}{2}(it, ju)]$$

where

$$\Phi_i, \Phi_j \in \text{ pair A}$$

and

$$\Phi_t, \Phi_u \in \text{ pair B}$$

---

[10]) Two perfect gases can inter-penetrate with no repulsion or attraction.

Using the separate-atom exponents for the GHOs (i.e. identical oxygen GHOs for both bond and lone pair) in an identical calculation gives, 2.484, 2.158 and 1.904 for the repulsions. Clearly the VSEPR model is rather stable with respect to changes in basis. In fact, the calculations reported in (28) (which are general VB calculations), and MO calculations show that the VSEPR rules are extremely 'robust' — they do not depend qualitatively on the details of the model of the molecular electronic structure. The MO model is a special case here since the localisation procedure is variationally arbitrary. In fact it is not too difficult to show that the most common localisation procedures have criteria for localisation which are much the same as the VSEPR rules: minimisation of exchange energy, maximisation of self-repulsion, maximisation of interorbital centroid distances etc. It is therefore more convincing to obtain a quantitative justification of these rules when a model is used which specifically includes energetic effects on localisation as the VB method does.

The GHO basis can therefore provide a localised, directional set of orbitals (hybrids) which do not have the principal qualitative disadvantage of the usual hybrid sets: they can be mutually orientated in any directions. What is more the directions taken up by the GHOs can be decided *variationally* and not by the unitary properties of a 'hybridisation matrix'. This conclusion means that the use of a GHO basis provides both a localised bonding picture and simultaneously a theoretical validation of the VSEPR rules. Thus, it is not necessary, for example, to *contrast* the 'hybrid method' and the 'VSEPR method' for molecular geometries (30): they are complementary.

Open Shell Systems

We have not mentioned 'open shells' of electrons in our general considerations but then we have not specifically mentioned 'closed shells' either. Certainly our *examples* are all 'closed shell' but this choice simply reflects our main area of interest: valence theory. The derivations and considerations of constraints in the opening sections are independent of the numbers of electrons involved in the system and, in particular, are independent of the magnetic properties of the molecules concerned simply because the 'spin variable' does not occur in our approximate Hamiltonian. Nevertheless, it is traditional to treat open and closed shells of electrons by separate techniques and it is of some interest to investigate the consequences of this dichotomy. The independent-electron model (UHF — no 'symmetry' constraints) is the simplest one to investigate: we give below an abbreviated discussion.

The energy expression for a single determinant Φ_0 of orthonormal spin-orbitals ψ_i is

$$E = \sum_i h_{ii} + \sum_{i>j} [(\psi_i \psi_i, \psi_j \psi_j) - (\psi_i \psi_j, \psi_i \psi_j)]$$

where

$$h_{ii} = \int dv \, \psi_i \hat{h} \, \psi_i.$$

Variations $\delta E^{(i)}$ in this energy expression due to variations $\delta \psi_i$ in the orbitals are zero if the orbitals are determined by the SCF MO equations

$$\hat{h}^F \psi_i = \epsilon_i \psi_i \tag{40}$$

where

$$\hat{h}^F = \hat{h} + \hat{J} - \hat{K}$$

where $\hat{J}$ and $\hat{K}$ are the usual 'Coulomb' and 'exchange' operators. An important property of these self-consistent orbitals is the vanishing of the 'single-excitation' CI matrix elements. If Φ_i^n is a single determinant formed by removing an electron from ortibal ψ_i and placing it in orbital ψ_n (unoccupied in the determinant Φ_0) then

$$\int dV \, \Phi_0 \hat{H} \, \Phi_i^n = h_{in}^F = \int dV \, \psi_i \hat{h}^F \, \psi_n = 0 \tag{41}$$

since the ψ_i can be chosen to diagonalise $\hat{h}^F$. This result (the Brillouin Theorem) is true independent of the number of electrons in Φ_0. If we place the spin-eigenfunction constraint on the single determinant on Φ_0 then this result is, in general lost: only in the closed-shell case can (41) be recovered. This means that for many properties including the most familiar 'spin' property (electron-nucleus contact hyperfine coupling), a configuration interaction problem must be solved in *addition* to the self-consistency procedure. The paradox is: insisting that an open-shell wave function be an eigenfunction of spin often results in a specific spin property (the contact term) being spuriously zero in the single determinant approximation. We are therefore content to remain at the UHF level for any number of electrons.

There is a more serious problem with open shells associated with our strong orthogonality constraint. This is not specifically associated with electron spin but open shells of electrons exposes the problem clearly. In the electron-pair model (with strong orthogonality) the analogue of the Brillouin theorem means that excitations or ionisations of single groups of electrons remain *localised* in the particular electron-pair (29). Thus, in the absence of spatial symmetry, unpaired electrons are localised by the strong orthogonality requirement. In general this behaviour is not found experimentally: the existence of the so-called transferred hyper-fine interactions shows that even unpaired electrons which are nominally localised (in e.g. d shells) are in fact often de-localised. The fact is that this phenomenon for open shells is symptomatic of a wider failing of the electron pair model. Excited and unpaired electrons are generally less tightly bound than closed-shell ground state electrons: they are more delocalised, in general. It is therefore not too surprising that our localised electron-pair model is not appropriate in this area. It is clearly necessary to take account of the more diffuse and delocalised nature of the distribution of 'excited electrons', that is to use a more general model than the electron-pair model; or, of course, fall back on the MO model which can be localised or delocalised at will.

These latter considerations clarify our position on the use of particular models of molecular electronic structure. The electron-pair model is not *absolutely* preferable to the MO model in all respects, that is the electron-pair model is not to be recommended *per se*, but is to be preferred in most systems consisting of ground states of saturated bonds.

It is to be expected that the optimum scale factors of orbitals which are 'singly occupied' will show different behaviour from the 'closed shell' orbitals. We have therefore performed an optimisation of the GHOs in the CHO radical in the UHF model. The results are summarised below.

The CHO radical is a σ-radical and the main qualitative feature of the comparison between Tables 2 and 3 is the tendency for the σ GHOs to be less contracted in CHO than is found in CH_2O. In particular, the sp^2 hybrid on carbon which nominally contains the unpaired electron is considerably expanded (exponent 1.5923 compared to 1.8660 in CH_2O). The hydrogen orbital and the σ_{OC} orbital are also noticeably expanded while the lone pairs on oxygen are largely unaffected.

Table 3. Optimised GHO exponents for the CHO radical[a])

σ_{CH}	σ^*	σ_{HC}	σ_{CO}	σ_{OC}	π_{CO}	π_{OC}	σ_{O1}	σ_{O2}
1.7214	1.5923	0.9933	1.7352	3.2229	1.4986	2.1185	2.2835	2.2860

[a]) Notation as Table 2; σ^* is the (sp^2) GHO on Carbon which nominally contains the odd electron: σ_{O1} and σ_{O2} are the two (inequivalent) lone pairs on oxygen.

Conclusions

Before summarising our findings it is perhaps as well to look briefly at the problem we set ourselves, the possible attitudes to the results of our investigations and the general areas in which we would particularly like to see clarification.

In the first place, we have been studying those parts of the theory of the electronic structure of molecules which bear on a theory of *valence*. That is we are not attempting to present a theory of molecular electronic structure, but an approximate theory of valence. The latter is but a small part of the former. In particular we are (in the main) concerned with localised systems of electrons in their ground states: the theory of groups of tightly bound electrons with, individually, only one or two bound states. This restriction excludes the treatment of molecular excited states (except in a nominal way as the excited states of our valence model) and any systems of mobile, weakly bound electrons. The characteristic de-localised structure of con-jugated systems of π electrons is of particular interest here as it is an intermediate case; a system of electrons intimately involved in the bonding but relatively weakly bound to the nuclear framework and with many bound states in close proximity. We might therefore expect that a careful study of π-electron systems to yield information about the changeover from a tightly-bound 'valence' situation to a more loosely bound 'excited-state' situation. In this context we might mention that the AO ex-pansion method itself is only really suited to the valence situation since the minimal AO basis functions reflect a tightly bound model.[11] Thus, the whole minimal basis AO expansion method is not suitable for excited states in general or (perhaps!) for systems of de-localised electrons.

Secondly, we might mention that there are two possible attitudes to a theory of valence based on the AO expansion method. The chemist uses the electron-pair model essentially as one of his axioms or, at least, as a good working hypothesis. This model is extremely familiar and useful; any *theory* of such a model of valence should be capable of providing at least some foundation for and extension of his qualitative concepts. To the physicist, however, the simplest molecules are quite complex many-particle systems and he would perhaps find it surprising if we are able to obtain *any* useful results from our coarse-grained minimal basis model in view of the complexity of the interactions involved. We must try to balance these views in any evalution we make.

Thirdly, any theory of valence should contain some elucidation of at least the following points:
  (i) why chemical bonding occurs: why a sharing of electrons between two (or more) attracting centres leads to a stabilisation;
  (ii) why electrons are shared in pairs: the overwhelming predominance of the electron pair bond;
  (iii) how molecular shapes are determined: the interactions between the bond pairs.

We can, unfortunately, only claim any success in areas (i) and (iii). The electron-pair model, far from emerging from our analysis, was chosen as a starting point – a particular case of Eq. (6). In fact, using an orbital-basis expansion technique and imposing the Pauli principle it is difficult to see how this impasse can be avoided: the electron-pair model is still an explicans not an explicandum.[12]

However, in areas (i) and (iii) we have been able to obtain some results. These results are qualitative and theoretical (as opposed to quantitative and numerical). We have been able, among other things, to provide a unified theoretical foundation for

---

[11] The free-electron model of π systems is the obvious extreme case.

[12] It might be thought that electron-spin 'coupling' provides an explanation of the characteristic 'two-ness' of electrons in molecules. This is not so; spin coupling is a kind of mnemonic device for the *formulation* of the electron-pair model, not an explanation of it.

two of the most satisfactory aspects of the existing theory of valence: Ruedenberg's analysis of the factors contributing to bond formation (*18*) and the VSEPR model for molecular geometries (*25*). Using the GHO minimal basis it is clear from the relatively small amount of numerical data available to date that the qualitative behaviour of the GHOs on being transferred from an atomic to a molecular environment is in substantial agreement with Ruedenberg's comprehensive and quantitative analysis of the bonding in H_2^+. The characteristic contractions of the GHOs involved in σ bonding and the relative invariance of the lone-pair orbitals both have immediate physical interpretation. The GHOs are not generated by a linear transformation of an ordinary AO set, and so are free from the 'directional constraint': their mutual directions are not determined in advance and so provide an ideal basis for the modelling of the electron pairs in the VSEPR model.

The fact that the GHOs, when optimised in a molecular environment, generate an atomic orbital basis which has just the symmetry of the molecule (i.e. not higher or lower symmetry) is extremely encouraging — it tends to reassure that the GHO basis does refelct the physics of the interactions adequately. Thus, *within the approximation scheme of a minimal GHO basis,* the 'symmetry dilemma' is actually a 'choice of basis dilemma' and can be satisfactorily solved by using near-optimum scale factors for the GHOs.

The GHO basis also has the property of validating the NDO family of approximations more satisfactorily than any other atomic orbital basis. This basis provides the justification for regarding the NDO schemes as rather more than convenient numerical procedures: using a GHO basis the NDO approximations are physically realistic. The non-orthogonalities in this atomic orbital basis are concentrated into pairwise — bonded GHOs and the orbital contraction effects ensure that these non-orthogonalities are at a minimum. These properties mean that the *'essential'* non-orthogonalities in the AO basis (which, overtly or covertly, lead to bound molecules) are retained in a chemically transparent way *and* the 'inessential' non-orthogonalities (between pairs of nonbonded GHOs) are minimised to such an extent that, in a semi-quantitative theory, they may be ignored.

The upshot of these conclusions is that the use of a minimal basis of GHOs provides the formal, conceptual and numerical foundation for a variety of the aspects of valence theory. In particular, the GHOs act as a theoretical foundation for a number of facets of the theory of valence, which are not normally considered to be intimately connected:

(i) The use of hybrid orbitals (independently of the model of molecular electronic structure used).
(ii) The orbital basis 'symmetry dilemma'.
(iii) The energetic analysis of chemical bonding.
(iv) The VSEPR model.

and

(v) The NDO family of approximations.

Although we have been at some pains to point out that the choice of orbital basis is formally independent of the choice of specific model of electronic structure (MO, VB etc.), many of the conclusions above can be reinforced by using the GHO basis together with an electron-pair model (or electron-pair plus de-localised groups where these exist). Just as the use of a basis of GHOs (with variationally-determined exponents) frees us from the unitary transformation identity in MO theory, so the use of a localised electron-pair model enables a non-arbitrary (variational) choice of localised model to be made. Together, these two approximations make near-optimal use of the variational degrees of freedom inherent in the minimal-basis description: the variation theorem enables the optimisation of non-linear parameters (GHO exponents) and linear parameters (matrix Schrödinger equation) to be carried through.

As we noted at the outset, we are not attempting to present a computational technique for the accurate calculation of molecular electronic structure: this is an essentially technical problem — the optimum series expansion solution of a multi-dimensional partial differential eigen problem. We are concerned here to present a model of valence which directly reflects the qualitative processes occurring on bond formation. This model is, of course, complementary to any powerful numerical techniques for the precise computation of molecular energies, but it is *not reducible* to such a technique. The powerful numerical methods have proved extremely difficult to interpret, in fact most of the conceptual framework of valence theory is based on rather simple models. For example, we have not mentioned electron correlation as such (although the electron-pair model includes correlation) because, in spite of the enormous amount of quantitative work in this area, there have been no concepts evolved. Indeed, since the very definition of electron correlation depends on a specific *model* of electronic structure (i.e. correlation is not a phenomenon) it is unlikely that such concepts will emerge. To take an obvious example, the fluorine molecule F_2 is not bound in the Hartree-Fock model: the very existence of a stable molecule is 'dependent on electron correlation'. Yet no-one has found it necessary to postulate a qualitatively new form of bond in F_2 (a 'correlation bond', say) because the bonding in F_2 is patently not different from the bonding in Cl_2.

The idea that a precise computation of a quantity is a substitute for, or even identical with, an explanation of that quantity has a long and dishonourable history. Starting with the Ptolomaic model of the solar system (which, initially, was in better quantitative agreement with observations than the helio-centric model) and coming to a sort of climax in modern field theories with infinite self-energies and subtracted-out divergences (yet in almost perfect agreement with experiment) this attitude has been a persistent thread in science. We wish to keep the scientific problems of explanation distinct from, but of course complementary to, the technical problems of precise numerical calculation. Thus, our minimal GHO basis cannot compete with an extended basis in quantitative energy calculations, but on the other hand the more comprehensive calculations are uninterpretable as theories of *valence*. An analogy may be helpful: the *explanation* of the existence of tides on the oceans consists of the simple fact of the moons gravitational pull. But a *calculation* of the height of any particular tide might involve an expansion of the moons gravitational field (in spherical harmonics, say) not to mention the effect of the sun, the topography of the coast line, the prevailing winds etc. etc. Notwithstanding all these latter considerations and their quantitative importance, it would be a brave man who, on the evidence that a high tide did not occur when the moon was overhead, asserted that the tides are *not* due to the moon's pull.

References

1. *Cook, D. B.:* Structures and Approximations for Electrons in Molecules. Chichester: Ellis Horwood 1977
2. *Löwdin, P.-O.,* see for example Research Report No. 40 of Quantum Theory Project, University of Florida (1963)
3. *Pople, J. A., Segal, G. A.:* J. Chem. Phys. *43*, S129 (1965)
4. *Cook, D. B.:* Theoret. Chim. Acta *40*, 297 (1975)
5. *Cook, D. B., Hollis, P. C., McWeeny, R.:* Molec. Phys. *13*, 573 (1967)
6. *Linderberg, J.:* Chem. Phys. Lett. *1*, 39 (1967)
7. *Löwdin, P.-O.:* J. Chem. Phys. *18*, 365 (1950)
8. *Kolos, W., Roothaan, C. C. J.:* Rev. Mod. Phys. *32*, 219 (1960)
9. *McWeeny, R.:* In: The New World of Quantum Chemistry. B. Pullman and R. G. Parr (Eds.) pp. 3–31. Dordrecht: Reidel 1976
10. *Parr, R. G.:* Quantum Theory of Molecular Electronic Structure. New York: Benjamin 1964
11. *Cook, D. B.:* Ab Initio Valence Calculations in Chemistry. London: Butterworths 1974
12. *Cook, D. B.:* Theoret. Chim. Acta (in press, 1977)
13. *Cooper, I. L., McWeeny, R.:* J. Chem. Phys. *49*, 3223 (1968)
14. *Löwdin, P.-O.:* Adv. Quant. Chem., *5*, 185 (1970)
15. *Kirkwood, E. F., Cook, D. B.:* Theoret. Chim. Acta *44*, 139 (1977)
16. *Roby, K. R.:* Chem. Phys. Lett. *12*, 579 (1972)
17. *Cook, D. B., Fowler, P. W.:* (submitted for publication 1977)
18. *Ruedenberg, K.:* In: Localisation and Delocalisation in Quantum Chemistry, Vol. 1. Dordrecht: Reidel 1975
19. *Fletcher, R.:* Report R 7125 of the Atomic Energy Research Establishment U. K.
20. *Primas, H.:* Classical Observables in Quantum Mechanics (preprint – to be published)
21. *Harris, F. E., Michels, H. H.:* Adv. Chem. Phys. *13*, 205 (1967)
22. See, for example, V. R. Saunders report RL-76-104 (ATMOL 3) Atomic Energy Research Establishment U.K.
23. *Huzina, A. S.:* J. Chem. Phys. *42*, 1293 (1965)
24. *Brown, R. D., Roby, K.:* Theoret. Chim. Acta *16*, 194 (1970)
25. *Gillespie, R. J., Nyholm, R. S.:* Quart. Rev. (Lond) *11*, 339 (1957)
26. *Walsh, A. D.:* J. Chem. Soc. (Lond) 2260 (1953)
27. *Allen, L. C.:* Theoret. Chim. Acta *24*, 117 (1972)
28. *Wilson, S., Gerratt, J.:* Molec. Phys. *30*, 789 (1975)
29. *McWeeny, R., Klessinger, M.:* J. Chem. Phys. *42*, 3343 (1965)
30. *Pimental, G. C., Spratley, R. D.:* Chemical Bonding Clarified Through Quantum Mechanics. San Francisco: Holden-Day 1969

Applications of the Angular Overlap Model

Derek W. Smith

School of Science, University of Waikato, Hamilton, New Zealand

Table of Contents

1. Introduction

The angular overlap model (AOM) attempts the calculation of orbital splittings in a partly-filled shell in terms of parameters which are directly related to σ-, π-, δ- etc. bonding. Thus defined, the AOM has been in use for some twenty years, although the term did not appear until 1965. There is still a widespread feeling that the AOM is distinguished more for its elegance and simplicity than for its practical utility in the solution of chemical problems. The theory has been expounded in numerous articles and monographs (1–32) and will not be discussed in detail here; useful introductory accounts will be found in Refs. 2–4 and 29. This article reviews the practical uses of the AOM in the interpretation of experimental observations, and presents a critical analysis of the successes and failures of the model.

We begin by distinguishing the various alternative formulations of the AOM which have appeared in the literature. The origins of the AOM can be seen in the models of *Yamatera* (5, 6) and *McClure* (7), who were mainly interested in the d–d spectra of octahedral Co(III) chromophores. The splittings of the octahedral T states in substituted systems can be expressed in terms of the parameters $\delta\sigma$ and $\delta\pi$, which are related to the one-electron splittings of the octahedral e_g and t_{2g} orbitals respectively, and are defined as the differences between the ligands X and Y in MX_nY_{6-n} with respect to their σ- and π-antibonding effects on the metal d-orbitals. Thus *McClure* (7) was able to construct a two-dimensional spectrochemical series, in which the relative σ- and π-antibonding effects of ligands could be listed separately.

The next step in the development of the AOM was the Ξ^2 model (8–11), which initially considered only σ-antibonding effects of ligands but can be extended to include the effects of π-overlap as well. The overlap integral S_{MX} between a metal d-(or f-)orbital and a ligand group orbital is expressed as the product of an angular part Ξ and a radial part S_{MX}^*, and it can be shown from semi-empirical MO theory that the relative σ-antibonding effects on the orbitals of the partly-filled d- or f-shell are proportional to Ξ^2. Since Ξ can be explicitly determined from the geometry of the complex, the relative energies of the orbitals can be expressed as multiples of the proportionality constant σ^*.

The term 'angular overlap model' was first used in 1965 to describe a more general treatment, which may also be called the e_λ model (12, 13). The relative energies of the orbitals in a partly-filled l-shell are expressed in terms of the parameters e_λ ($\lambda = \sigma$, π, etc.), whose coefficients can be calculated from the geometry of the system. The e_λ model was developed from perturbation theory, but is equivalent to the Ξ^2 model when only σ-overlap is considered, and to the *Yamatera-McClure* model for orthoaxial chromophores with linear ligators. The notation e_λ' is often used, where:

$$e_\lambda' = e_\lambda - e_\delta \qquad (\lambda = \sigma, \pi).$$

Unless δ-overlap is ever significant, this notation is hardly necessary.

If the ligands are not linear ligators, two e_π parameters are needed, e_π ($\parallel$) and e_π ($\perp$). It is sometimes possible, by inspection of the ligand, to set one or both of these equal to zero. Since e_λ is deemed to be proportional to the square of the appropriate overlap integral, the radial dependence of e_λ can be found by calculating the overlap integral S_λ over a range of internuclear distances.

The e_λ model has been elaborated in a number of ways. An electrostatic perturbation was added (33) to account for band splittings in the $d-d$ spectra of tetragonal copper(II) ammine complexes where the simple AOM predicted accidental degeneracy; the merits of this refinement will be discussed in 2.5.1. Another development has been the introduction of $d-s$ and $d-p$ mixing, which is apparently necessary to account for the $d-d$ spectra of chlorocuprates(II) (34). This requires the additional parameters e_{ds}, $e_{dp\sigma}$ and $e_{dp\pi}$.

Burdett (35–38) has extended the AOM by the introduction of a quartic term in the expansion of the perturbation determinant as a power series in the overlap integral S_λ. In the conventional AOM, only the quadratic term (proportional to S_λ^2) is considered. In closed-shell systems, the sum of the energies of the relevant orbitals is independent of angular variations in the molecular geometry if only the quadratic term is used. This is no longer true if the quartic term is included, and it is possible to rationalise many stereochemical observations.

Other models, which have been developed independently, have features in common with the AOM. Kettle and his co-workers (39–42) have discussed molecular geometry using a perturbation treatment which places the relative energies of MO's in terms of squared overlap integrals, and the 'point bond' model of Howald and Keeton (43) is essentially equivalent to the e_λ model.

2. d–d Spectra

Most applications of the AOM concern the interpretation of $d-d$ spectra. This Section is divided into several areas of interest in which the AOM has been extensively used. Emphasis is placed on work which has attempted the extraction of explicit AOM parameters, rather than on the use of the model in a purely qualitative way.

2.1 Six-Coordinate Systems with Ground States of Nondegenerate Octahedral Parentage

Here we are concerned with octahedral complexes of Cr(III), Ni(II) and low-spin Co(III). The AOM has been extensively applied to the analysis of the $d-d$ spectra of subsituted chromophores MX_nY_{6-n}, particularly trans-MX_4Y_2. The function of the model here is to parameterise the one-electron matrix elements which represent the splittings of the central-ion d-orbitals in the noncubic ligand field. The parameters are deemed to be transferable from one system to another, and their evaluation should make it possible to separate the σ- and π-antibonding (and, perhaps, π-bonding) effects of various ligands. Both the *Yamatera-McClure* and e_λ formulations of the AOM have been used, as well as the additive and nonadditive crystal field models, and a great variety of approximations and assumptions have been invoked. A meaningful analysis depends first on the successful resolution of closely-spaced, overlapping, broad absorption bands and their correct assignment. In some cases, solution spectra have been subjected to the dubious procedure of Gaussian analysis to locate the maxima. Single crystal polarized spectra, measured if possible at low temperatures, afford better resolution and at least some of the bands can usually be unequivocally assigned, but only a limited amount of such data is available.

The energy matrices contain one-electron terms (which can be written down in terms of the AOM e_λ parameters) and two-electron terms, expressed as multiples of the Racah parameters B and C. Values of the e_λ and Racah parameters which provide the best fit to the experimental data are then found. Most work has been done on the tetragonal (D_{4h}) chromophores MN_4X_2, where the N atoms (equatorial) are provided by amine ligands. Only three AOM parameters can be determined since there are only three independent orbital splitting parameters; $e_\sigma(N)$, $e_\sigma(X)$ and $e_\pi(X)$ can be found if $e_\pi(N)$ is taken to be zero, saturated amines having no orbitals available for π-overlap. In some papers, only the spin-allowed bands have been used in the analysis; for d^3 and d^8 systems, this obviates the need to consider the Racah parameter C. Where the spin-forbidden bands have been included, C has sometimes been allowed to find a value which best fits the experimental data, along with the other parameters; others have assumed a fixed value of the ratio B/C, such as 0.25. The treatment of interelectron repulsion introduces some uncertainty into the orbital splitting parameters. Although it is well-known that the d–d spectra of O_h chromophores cannot be perfectly fitted within a model which allows only one value of B *(1–3, 44, 45)*, this assumption is nevertheless invariably made in dealing with noncubic systems, otherwise far too many parameters would be needed.

Some authors have sought to simplify the analysis by wholly or partly neglecting configuration interaction. This approximation introduces large errors for a small saving in labour. Much more acceptable is the usual neglect of spin-orbit coupling, although this could be important in some Ni(II) systems.

2.1.1 Chromium(III) Systems

Chromophores of the type $Cr(III)N_4X_2$ (D_{4h}) have been rather thoroughly studied within the AOM *(46–58)*. We shall concentrate on the most recent work *(58)*, in which an excellent review of the subject will be found. In O_h symmetry, the ground state of a d^3 system is $^4A_{2g}$, with quartet excited states $^4T_{2g}$, $^4T_{1g}(F)$ and $^4T_{1g}(P)$. In D_{4h}, these become:

$$^4A_{2g} \rightarrow {}^4B_{1g}$$
$$^4T_{2g} \rightarrow {}^4B_{2g} + {}^4E_g$$
$$^4T_{1g} \rightarrow {}^4A_{2g} + {}^4E_g$$

Thus the six spin-allowed transitions from the ground state $^4B_{1g}$ have energies which are given by the roots of a 3 × 3 determinant (E_g), a 2 × 2 determinant (A_{2g}) and a simple linear equation for B_{2g}. In practice, only four transitions are observable:

$$^4B_{1g} \rightarrow {}^4B_{2g}$$
$$\rightarrow {}^4E_g(a)$$
$$\rightarrow {}^4E_g(b)$$
$$\rightarrow {}^4A_{2g}(a)$$

The first two are usually well-resolved, even in solution, but the second pair are often rather close together. Assignment has been aided by the single crystal polarized spectra (52, 53, 56) of $[Cr(en)_2 X_2]^+$ (X = F, Cl, Br). From the four experimental quantities we can obtain the four theoretical parameters $e_\sigma(N)$, $e_\sigma(X)$, $e_\pi(X)$ and B. In the most thorough study (58), the AOM parameters have been found for NH_3, pyridine, F^-, Cl^- and Br^-. Pyridine is a non-linear ligator with the capability of π-overlap. By assuming that the $e_\lambda(X)$ parameters in $[Cr(py)_4 X_2]^+$ are the same as in $[Cr(NH_3)_4 X_2]^+$, it was found that pyridine is a π-acceptor towards Cr(III) (with a negative e_π) and that the angle between the plane of a pyridine molecule and the plane of the four equatorial N atoms is about 38°. Angular overlap parameters for various ligands in Cr(III) systems are collected in Table 1. The parameters for the non-linear ligators H_2O and OH^- are rather uncertain. The twodimensional spectro-chemical series for Cr(III) is found to be:

$$e_\sigma: \quad I^- < Br^- < Cl^- < py < RNH_2 < NH_3 < H_2O \sim F^- \sim en < OH^-$$
$$e_\pi: \quad py < NH_3 \simeq en < I^- < Br^- < Cl^- < H_2O < F^- < OH^-.$$

These series do not indicate the relative amounts of σ- and π-donation by various ligands; they reflect the effects of σ- and π-overlap on the energies of the central-ion d-orbitals. It has been argued (54) that the AOM parameters may be best interpreted within an ionic model, both σ- and π-electrons on the ligands contributing to destabilization of the metal d-orbitals; it is further argued that simultaneous σ- and π-donation by a ligand like F^- is unlikely. This reasoning has been challenged (57) on the grounds that the electrostatic perturbation produced d-orbital splittings in the opposite sense to those observed experimentally, and that only a little covalency is needed to account for the splittings. Indeed, even in the limit of zero covalency,

Table 1. Angular overlap parameters e_λ in tetragonal Cr(III) systems, taken from Refs. 57 and 58. All energies are in kK

Ligand	e_σ	e_π
NH_3	7.0	0
C_5H_5N	5.8	-1.0
$NH_2(CH_2)_2NH_2$	7.3	0
$NH_2(CH_2)_3NH_2$	7.4	0
$NH_2CH(CH_3)CH_2NH_2$	7.3	0
CH_3NH_2	6.7	0
$C_2H_5NH_2$	6.7	0
$C_3H_7NH_2$	6.6	0
F^-	7.4	1.7
Cl^-	5.5	0.9
Br^-	4.9	0.6
I^-	4.3	0.6
OH^-	9.0	2.0
H_2O	7.5	1.4
CH_3COO^-	7.0	0.2–1.0

metal-ligand overlap can lead to substantial d-orbital splitting (59). The relative magnitudes of e_σ and e_π for the halides are consistent with the view that metal-ligand overlap is responsible for the splitting. The ratio (e_π/e_σ) decreases with increasing atomic number of the halogen, and it can be argued that this is consistent with the decreasing ratio of the squared overlap integrals $(S_\pi/S_\sigma)^2$ as the internuclear distance is increased. However, (e_π/e_σ) always seems to be smaller than $(S_\pi/S_\sigma)^2$. If we recognise that both ns- and np-orbitals on the ligands contribute to the σ-bonding, then e_σ must be the sum of two components, $e_{\sigma s}$ and $e_{\sigma p}$.

The ratio $(e_{\sigma s}/e_{\sigma p})$ is expected to decrease in magnitude as the energy separation between the ns- and np-orbitals increases. Hartree-Fock calculations (60) show that the $ns-np$ separation for F, Cl and Br is respectively 0.84, 0.57 and 0.54 a.u.; for F^-, Cl^- and Br^-, these increase to 0.89, 0.58 and 0.55 a.u. respectively. Thus we expect $(e_{\sigma s}/e_{\sigma p})$ to increase with increasing atomic number of the halogen, and hence (e_π/e_σ) should decrease as interaction with the ligand ns-orbitals increases.

2.1.2. Cobalt(III) Systems

The situation in low-spin Co(III)-containing chromophores such as CoN_5X and CoN_4X_2 is much less favourable than for the Cr(III) analogues. The octahedral ground state is $^1A_{1g}$, with singlet excited states $^1T_{1g}$ and $^1T_{2g}$; the latter are split into two components each in D_{4h} or C_{4v}. The components of the octahedral $^1T_{1g}$ state are usually observed experimentally, but the splitting of the $^1T_{2g}$ state has rarely, if ever, been resolved. Spin-forbidden transitions to triplet states must be included in the analysis. Even if the latter are neglected, it is still necessary to introduce the Racah parameter C, as well as B. Some qualitative features of the $d-d$ spectra of CoN_5X, CoN_4X_2 (cis- and trans-) and CoN_3X_3 chromophores can be rationalised by application of the AOM (5–7, 12). However, it is difficult to extract useful AOM parameters; a brave attempt was made by *Wentworth* and *Piper* (61).

Schmidtke (62) has shown how the AOM can be used to assign the geometrical isomers of bis(L-hitidinato)cobalt(III) by parameterisation of the $^1T_{1g}$ splitting, without explicitly evaluating the AOM parameters.

2.1.3. Nickel(II) Systems

Octahedral Ni(II) complexes have the ground state $^3A_{2g}$, and the energy matrices for O_h and its subgroups are the same as those for Cr(III), apart from the different spin multiplicities. Since the $d-d$ bands of Ni(II) systems lie at lower energies than those of analogous Cr(III) systems, all the spin-allowed bands are usually observed. Thus, in principle, we should have more data to work with. Unfortunately, the observed splittings of the octahedral T states are usually rather small, especially for $^3T_{1g}(P)$. Thus the calculated splitting of the octahedral t_{2g} orbitals is always found to be small, and is probably poorly-determined. The values of $e_\pi(X)$ found for NiN_4X_2 chromophores are likely to be rather uncertain. In $[Ni(NH_3)_4(NCS)_2]$, for example, e_π for NCS$^-$ is found to be 0.1 kK, but in $[Ni(en)_2(NCS)_2]$ this parameter is apparently equal to -0.4 kK. If these figures are to be believed, NCS$^-$ is a π-donor in one compound and a π-acceptor in the other, which seems to be highly unlikely. For $[Ni(NH_3)_4(NO_2)_2]$ and $[Ni(en)_2(NO_2)_2]$, e_π for NO$_2^-$ is consistently slightly negative, ca. -0.1 kK, but the significance of this must be regarded as doubtful (48, 63).

In $[Ni(py)_4X_2]$, pyridine apparently has a larger e_π than Cl$^-$ or Br$^-$ (48, 64–67), suggesting that pyridine is a π-donor towards Ni(II). However, it will be recalled from 2.1.1. that pyridine is apparently a π-acceptor in Cr(III) complexes (58). While it is quite possible that pyridine could have e_π values of opposite sign with different metals, it is surely more likely to function as a π-acceptor towards Ni(II) than towards Cr(III). We shall comment further on this discrepancy in Section 7.

It seems that e_σ for pyridine in $[Ni(py)_4X_2]$ (X = Cl, Br, I) is somewhat dependent on the axial ligands; the same is true for $[Ni(en)_2X_2]$ and $[Ni(NH_3)_4X_2]$; the variation of $e_\sigma(N)$ is in accordance with its proportionality to S_σ^2 (63). *Hitchman* (66) has made use of the assumed proportionality of (e_π/e_σ) for halides to $(S_\pi/S_\sigma)^2$ to fix the relative values of $e_\pi(X)$ and $e_\sigma(X)$ in NiN_4X_2 chromophores, making it possible to determine $e_\pi(N)$. This was always found to be positive for pyridine, pyrazole and thiourea. However, his values of (e_π/e_σ) for the axial ligands may be too large, for the reasons discussed above (2.1.1.).

The two-dimensional spectrochemical series for Ni(II) has yet to be established with much certainty, since, as we have already discussed, the e_π parameters are poorly-determined. The series for σ-bonding is better known, and runs as follows:

2-methylimidazole $>$ pyridine $>$ pyrazole $>$ NO_2^- $>$ NCS^- $>$ NH_3 $>$ Cl^- $>$ Br^- $>$ I^-.

The relative positions of pyridine and ammonia are reversed compared with the series for Cr(III); this is also true for the conventional (one-dimensional) spectrochemical series. The latter observation could be attributed to the pyridine molecule acting as a stronger π-donor towards Cr(III) than towards Ni(II), contrary to the conclusions of *Glerup et al.* (58), who found pyridine to be a π-acceptor in Cr(III) systems. However, it seems to be fairly clear from the spectra of [Ni(py)$_4$X$_2$] (48, 64–67), as discussed above, that the d_{xy} orbital lies higher in energy than $d_{xz, yz}$. In [Ni(py)$_4$I$_2$], the angle between the plane of a pyridine ring and the NiN$_4$ equatorial plane is close to 45°, so that pyridine is effectively a linear ligator (68). Hence, e_π for pyridine must be greater than e_π for halide. This paradoxical situation could suggest that the assumption of transferability of $e_\pi(X)$ parameters, invoked by *Glerup et al.* (58) in order to determine e_π for pyridine, is not justified (see also Section 7).

2.2 Six-Coordinate Iron(II) Systems

In *trans*-Fe(II)N$_4$X$_2$ chromophores, the ground state may be ^5E_g or $^5B_{2g}$, assuming that they are high-spin. The nondegenerate ground state should arise if the equatorial ligands are poorer π-donors than the axial ligands. Most spectroscopic measurements have been made on systems where the N-donors are aromatic ligands such as pyridine and phenanthroline (48, 69); saturated amine complexes of Fe(II) are less accessible. These ligands apparently produce a $^5B_{2g}$ ground state, according to Mössbauer studies (70), but the results are not unambiguous and in any case spin-orbit coupling will mix the $^5B_{2g}$ and ^5E_g states. The two observed d–d transitions are likely to be to the excited states $^5A_{1g}$ and $^5B_{1g}$, which arise from splitting of the octahedral ^5E_g state. The assignment can be made on the basis of the assumption that $e_\sigma(N) > e_\sigma(X)$, so that $E(^5B_{1g}) > E(^5A_{1g})$. There are insufficient data from which to extract explicit AOM parameters, but the McClure parameter $\delta\sigma$ can be found – in magnitude, if not in sign – from the splitting of the $^5B_{1g}$ and $^5A_{1g}$ states: the sign is uncertain for [Fe(py)$_4$(NCS)$_2$] and [Fe(py)$_4$(NCSe)$_2$] (48, 69).

Compounds such as Fe(py)$_2$Cl$_2$ contain polymeric chains, with *trans*-FeX$_4$N$_2$ chromophores. However, the symmetry is now D$_{2h}$ and d-orbital mixing may complicate the issue (70). We might expect a ^5E_g ground state (ignoring the rhombic distortion) but this could be further split by the Jahn-Teller effect. The spectra of such systems can be partially interpreted (48, 69) on the basis of a ^5E_g state in D$_{4h}$, with the $^5A_{1g}$ state higher in energy than $^5B_{1g}$. It must be noted, however, that even Fe(II)X$_6$ chromophores, with apparently little static distortion, can exhibit band splittings which may be attributed to Jahn-Teller distortions in the ^5E_g excited state. The extraction of useful AOM parameters from the d–d spectra of tetragonal Fe(II) complexes remains a difficult problem.

2.3 Five-Coordinate Systems (other than d^9)

The development of the AOM in the mid-1960's coincided with an explosion of activity in the study of five-coordinate systems; unfortunately, many of these have symmetries which are rather lower than the idealised D_{3h} (trigonal bipyramid) or C_{4v} (square pyramid). Even so, some useful information has arisen from the AOM.

2.3.1 The Pentachlorovanadate(IV) Ion

A cautionary tale now unfolds. The vibrational spectrum of PCl_4VCl_5 (71) was interpreted in terms of a trigonal bipyramidal VCl_5^- ion, possibly with some distortion. Analysis of the $d-d$ spectrum using the AOM (72) suggested that square pyramidal coordination geometry was more probable. A trigonal bipyramidal configuration was rejected on the grounds that no reasonable value of the ratio (e_π/e_σ) would fit the spectrum in D_{3h} symmetry. However, an analysis using simple crystal field theory (73) led to good agreement with the experimental spectrum if the anion were a somewhat distorted trigonal bipyramid. The crystal structure of PCl_4VCl_5 has now been published (74); the compound does indeed contain discrete VCl_5^- ions, whose geometry is very close to that predicted from the crystal field calculations (73). This would appear to be a triumph for the crystal field model and a disaster for the angular overlap model. The failure of the AOM in this case may arise, in part, from neglect of $d-s$ and $d-p$ mixing, which is likely to be particularly important for metals at the beginning of a transition series where the metal nd-orbitals lie quite close in energy to the $(n + 1)s$- and $(n + 1)p$-orbitals. For transition metal halide complexes, it seems that not only do we have to consider $d-s$ and $d-p$ mixing, but moreover we cannot determine the ratio (e_π/e_σ) from the squared overlap integrals, considering only ligand np-orbitals. To extract all the relevant AOM parameters for VCl_5^-, we would need more spectroscopic data for a range of chlorovanadate(IV) systems.

2.3.2 Cobalt(II) Systems

Five-coordination constitutes an important part of cobalt(II) chemistry; both the crystal field and angular overlap models have been used in the analysis of their $d-d$ spectra. Earlier work in this area has been reviewed previously (75).

  Although Co(II) forms both square pyramidal and trigonal bipyramidal complexes, most spectroscopic studies have been performed on the latter variety. *Norgett* and *Venanzi* (76) analysed the $d-d$ spectrum of $[CoCl(QP)]^+$ where QP is *tris*-(o-diphenylphosphinophenyl) phosphine. This has a CoP_4Cl chromophore, with distorted C_{3v} symmetry. The ground state is a spin doublet, and three spin-allowed $d-d$ transitions were observed. Only two AOM parameters were defined, $e_\sigma(eq)$ and $e_\sigma(ax)$.

Thus no information regarding the possible effects of π-bonding was forthcoming. The optimum value of the ratio $e_\sigma(\text{eq})/e_\sigma(\text{ax})$ was 0.8, consistent with the crystallographic data.

Other studies of trigonal bipyramidal Co(II) complexes have concerned high-spin species, with quartet ground states. *Bertini et al.* (*77*) have measured the polarized crystal spectrum of the trigonal bipyramidal CoO_5 chromophore in pentakis(2-picoline N-oxide)cobalt(II) perchlorate. The spectrum was assigned in C_{2v}, rather than D_{3h}. Energy level diagrams showing the variation of the $d-d$ transition energies with the ratio (e_σ/e_π) were constructed, but explicit AOM parameters were not obtained. It was concluded that e_σ/e_π must be either less than -1 or greater than 4; the latter figure is much more probable, so that the N-oxide donor is evidently a π-donor towards Co(II).

Bertini et al. (*78*) have also studied trigonal bipyramidal $CoNS_3Br$, $CoNP_3Br$ and $CoNN_3Br$ chromophores, and assigned their spectra in C_{3v}. In order to reduce the number of AOM parameters, the usual approximations were made; $e_\pi(N)$ was set equal to zero for saturated nitrogen donors, and the ratio (e_π/e_σ) for Br^- was estimated from the squared overlap integrals, assuming that only Br 4p-orbitals are involved. The e_λ parameters for a particular ligand were assumed to vary with the internuclear distance in accordance with the square of the appropriate overlap integral S_λ^2. These assumptions led to very reasonable values for the AOM parameters. The S-donor ligand tris(2-t-butylthioethyl)amine apparently behaves as a π-acceptor with Co(II), as does the P-donor tris(2-diphenylphosphinoethyl)amine. Moreover, the nephelauxetic ratio β required to fit the spectra of the $CoNS_3Br$ and $CoNP_3Br$ systems was substantially lower than that required for $CoNN_3Br$, as expected. The sequence $Br > N > S \sim P$ was found for both e_σ and e_π parameters; the position of bromide in the σ-spectrochemical series relative to N is surprising.

Ligand field matrices for $C_{3v} d^7$ systems, including spin-orbit coupling and in a form which can be readily correlated with AOM parameters, have been published (*79*).

Low-spin Co(II) complexes with Schiff base ligands such as N,N'-ethylenebis(salicyladiminate) (salen) have been extensively studied, but there has been much controversy concerning the ground state and the interpretation of the optical and magnetic properties. [Co(salen)] is square-coplanar and four-coordinate, while its dimer $[Co_2(\text{salen})_2]$ and the pyridine adduct [Co(salen)(py)] are five-coordinate and square-pyramidal. *Hitchman* has recently presented a detailed discussion of the electronic properties of these systems (*80*). Measurements on analogous copper(II) systems were used to assist the interpretation of the data for the Co(II) complexes. Hitchman concludes that the ground state in [Co(salen)] is $^2A_2(d_{yz})$, while the five-coordinate systems have $^2A_1(d_z2)$ ground states. The AOM parameter e_σ for salen takes the high value of about 11 kK, with $e_\pi(\|)$ about 4 kK and $e_\pi(\perp)$ about 6 kK. For [Co(salen)(py)], e_σ for pyridine was found to be about 7 kK and e_π about 1 kK, assuming pyridine to be effectively a linear ligator. Thus pyridine seems to be a π-donor towards Co(II).

2.3.3 Nickel(II) Systems

Surprisingly little use has been made of the AOM in the analysis of the $d-d$ spectra of five-coordinate Ni(II) complexes. Apart from some rather qualitative studies, the only work which has attempted to extract AOM parameters has been performed (66) on the square-pyramidal systems [Ni(2-meim)$_4$X] X, where 2-meim is 2-methylimidazole and X is chloride or bromide. Thus we have an NiN$_4$X chromophore, with X axial. The compounds are paramagnetic; the ground state is a spin triplet. The single crystal polarized spectrum was assigned in C$_{4v}$ symmetry; both spin-allowed and spin-forbidden transitions were included in the analysis. To reduce the number of AOM parameters, the ratio (e_π/e_σ) was estimated from the squared overlap integrals, assuming, of course, that only halogen np-orbitals are involved in the bonding. Thus the spectra could be fitted to the AOM parameters e_σ(eq), e_π(eq) and e_π(X), as well as the Racah parameters B and C. The AOM parameters thus found were consistent with those obtained for other tetragonal Ni(II) systems, with 2-methylimidazole functioning as a π-donor.

2.4 Triatomic Dichlorides of the Transition Metals

Molecular dichlorides MCl$_2$ of the $3\,d$ transition series elements can be observed in the gas phase or in inert matrices; they are presumed to be linear, although complete structural data are lacking. The electronic spectra of these molecules have aroused great interest and controversy (81–87). The earlier crystal field analyses suffered from inadequate treatment of the interelectron repulsion parameters (81, 82). The present author (85, 86) attempted to analyse these spectra in terms of AOM parameters (with an additional crystal field perturbation) which were consistent (after allowing for variations in the internuclear distances) with the parameters required to fit the spectra of other metal chloride complexes such as MCl$_4^{2-}$. These constraints led to some assignments which differed from those previously proposed (82), particularly in the case of CuCl$_2$. In a linear MCl$_2$ molecule, qualitative arguments suggest the energy sequence $\sigma(z^2) > \pi\,(xz, yz) > \delta\,(xy, x^2 - y^2)$ for the ligand field orbitals, so that for CuCl$_2$ we expect to see two $d-d$ transitions, $^2\Sigma \rightarrow {}^2\Pi$ and $^2\Sigma \rightarrow {}^2\Delta$. Two transitions are in fact observed, at 9 kK and 19 kK, but the latter is too intense to be of $d-d$ origin (84). Angular overlap parameters consistent with those of other chlorocuprates(II) led to the assignment of the 9 kK band to $^2\Sigma \rightarrow {}^2\Pi$, with the $^2\Sigma \rightarrow {}^2\Delta$ transition at ca. 16 kK where it would be obscured by the intense charge transfer band at 19 kK (85, 86). *Lever* and *Hollebone* (87) challenged this assignment, on the grounds that the orbital splittings in CuCl$_2$ should not be very different from those in the other dihalides, where the overall spread of the d-orbital energies is in the range 7–11 kK. We are now inclined to agree with their proposal (originally suggested by *DeKock* and *Gruen* (82)) that the 9 kK band is $^2\Sigma \rightarrow {}^2\Delta$, with $^2\Sigma \rightarrow {}^2\Pi$ lying at *ca.* 4 kK where it has not been observed for instrumental reasons. The fault in our original assignment apparently lies in our insistence that the parameters used to fit the spectra of MCl_2 should be consistent with those found for other metal halide complexes; the latter were evaluated without regard for $d-s$ and $d-p$ mixing, which we now know to be important in halide complexes (34). Moreover, we now recognise that in chlorocuprates(II), as well as in other species, the ligand ns-orbitals apparently contribute to e_σ so that our assumptions concerning the ratio (e_π/e_σ) are invalid. It is still possible, however, to rationalise the trends in the orbital splitting parameters for MCl_2 along the series. The orbital energies can be expressed as:

$$E(xy, x^2 - y^2) = 0$$
$$E(xz, yz) = 2 e_\pi$$
$$E(z^2) = 2 e_\sigma - 4 e_{ds}$$

We define the orbital splitting parameters $\Delta_1 = E(z^2) - E(xz, yz)$ and $\Delta_2 = E(z^2) - E(xy, x^2 - y^2)$, so that:

$$\Delta_1 = 2 e_\sigma - 2 e_\pi - 4 e_{ds}$$
$$\Delta_2 = 2 e_\sigma - 4 e_{ds}$$

How should these vary with M across the transition series? We suspect that the ratio (e_π/e_σ) should decrease as we cross from left to right, for two reasons. First, the metal-ligand π-overlap integrals decrease relative to the σ-overlap integrals across the series (86). Second, it may be argued that the contribution of the ligand ns-orbitals to e_σ should increase as we pass along the series, as the d-orbitals fall in energy and approach the ligand ns-orbitals more closely in energy. We also suspect that e_{ds}, the parameter which measures the extent of $d-s$ mixing, should decrease as we go across the series, as the $4s-3d$ energy separation increases. Thus we expect both Δ_1 and Δ_2 to increase across the series. The data in Table 2 are in broad agreement with these expectations, although the parameters for $CrCl_2$ are a little out of line. This may be explained if we argue, following *Lever* and *Hollebone* (87), that occupancy of the σ-antibonding orbital a_1' (d_{z^2}) has the effect of reducing e_σ and increasing e_π; all the dichlorides save $CrCl_2$ have one electron in the a_1' orbital in their ground states. We suggest that the AOM parameters in the MCl_2 molecules are approximately: $e_\sigma = 8-9$ kK, $e_\pi = 2-3$ kK and $e_{ds} = $ ca. 2 kK. Extrapolation of these parameters to

Table 2. Orbital splitting parameters for MCl_2 molecules. Data are taken from Ref. 82. No data are available for $MnCl_2$. All energies are in kK

	Cr	Mn	Fe	Co	Ni	Cu
Δ_1	5.4	–	2.5	3.2	3.4	ca. 4
Δ_2	9.0	–	7.1	8.3	9.6	9.0

bond lengths appropriate to typical octahedral and tetrahedral complexes produces reasonable values of the cubic splitting parameters.

Finally, it should be pointed out that the MCl_2 molecules may not all be perfectly linear (88).

2.5 Copper(II) Systems

The $d-d$ spectra of copper(II) compounds have provided a fruitful field for practitioners of the AOM. A wealth of structural data is available, and a rich variety of coordination geometries has been revealed. If we can make allowance for the dependence of the AOM parameters on the internuclear distance, we are provided with excellent opportunities to test the validity of AOM parameters over a range of related systems. However, the progress of such studies over the years has illustrated the fact that a simple model may be very successful in explaining a limited amount of dubious experimental data; as more crystal structures appear and as better spectroscopic data become available, the simple model may require considerable refurbishment, perhaps to the extent that it loses some of its appeal and utility.

2.5.1 CuN_4X_2 Chromophores

Tetragonal copper(II) ammine complexes have been considered to contain CuN_4X_2 chromophores, or CuN_4 square coplanar chromophores, since the axial ligands X are rather distant from the metal. There is little doubt, however, that the axial ligands exert a profound influence on the electronic properties of the system (89) and must be explicitly considered in any angular overlap analysis. The ground state is unequivocally $^2B_{1g}$, assuming D_{4h} symmetry, with excited states $^2A_{1g}$, $^2B_{2g}$ and ^2E_g, arising from electron transfer to the b_{1g} orbital $(d_{x^2-y^2})$ from $a_{1g}(d_{z^2})$, $b_{2g}(d_{xy})$ and $e_g(d_{xz,yz})$ respectively. The relative energies of the excited states are found, from polarized crystal spectra (90–92), to be $^2A_{1g} < ^2B_{2g} < ^2E_g$, although the $^2B_{2g}$ state is rarely observed and in the case of $Na_4[Cu(NH_3)_4][Cu(S_2O_3)_2] \cdot H_2O$, which represents the closest approach to a genuine CuN_4 square coplanar chromophore (93), there is a suspicion that the $^2B_{2g}$ state may lie higher in energy than ^2E_g (94, 95). In this compound, the $^2A_{1g}$ state is observed at 18.4 kK and the ^2E_g state at 19.2 kK. In tetragonal copper(II) ammine compounds of lower tetragonality, $^2A_{1g}$ is usually found at ca. 13 kK, $^2B_{2g}$ (when observed) at ca. 15 kK and ^2E_g at ca. 18 kK. The $^2B_{1g} \rightarrow {}^2A_{1g}$ transition energy is evidently very sensitive to the axial field.

From the AOM point of view, there are two rather puzzling features in these observations. The first is the fact that the $^2B_{2g}$ and ^2E_g states are not accidentally degenerate, which would be predicted if the equatorial ligands are assumed to have no π-overlap capability. In CuN_4X_2 chromophores, the distant axial ligands might have sufficient π-overlap to remove the degeneracy, but this would be in the sense $e_g > b_{2g}$ and would surely be very small. The second problem concerns the unexpectedly high energy of the $^2A_{1g}$ state; in the CuN_4 chromophore, with only σ-bonding, the $^2B_{1g} \rightarrow {}^2A_{1g}$ transition energy should be equal to $2e_\sigma$, while the $^2B_{1g} \rightarrow {}^2E_g$ transition energy is $3e_\sigma$. Thus the two transition energies should be in the ratio 3/2. This ratio should increase as axial ligands are added. In fact, the ratio is always less than 3/2. The present author suggested (33) that both these difficulties could be overcome by adding a crystal field perturbation, calculating the electrostatic parameters from realistic wave functions. This has the effect of splitting the b_{2g} and e_g orbitals in the correct sense, and also stabilizes the a_{1g} orbital relative to the others. If this correction is at all chemically meaningful, we should expect that the splitting of the b_{2g} and e_g orbitals should increase (with b_{2g} always higher than e_g) as the axial ligands are removed to infinity. In fact, the $^2B_{2g}$ and ^2E_g states cannot be distinguished in $Na_4[Cu(NH_3)_4][Cu(S_2O_3)_2] \cdot H_2O$, the most tetragonally-distorted system known, and the $^2B_{2g}$ state may actually lie higher than ^2E_g, according to e.s.r. results (93–95). We therefore feel forced to abandon the use of the electrostatic correction; Gerloch and Slade (3) have raised pertinent objections to this procedure. It has been shown (94, 95) that the relative energies of the $^2B_{2g}$ and ^2E_g states in these systems can be explained if two-electron terms are introduced into the expressions for the transition energies, to take account of symmetry-restricted covalence. In more complex systems, where all the ligands are capable of π-overlap, this treatment becomes rather cumbersome. We are thus unwilling to abandon the simple AOM, but we must recognise that in copper ammine complexes we cannot account for the relative energies of the $^2B_{2g}$ and ^2E_g states using a simple one-electron model. It may be noted that this phenomenon is likely to be important only in systems with a strong axial distortion, where $e_\sigma(ax) \ll e_\sigma(eq)$, and is probably of little significance in d^3 and d^8 systems, where no Jahn-Teller effect is operative.

If we ignore the $^2B_{2g}$ state (which is rarely observed anyway), the relative energies of the $^2A_{1g}$ and ^2E_g states cannot be accounted for unless we make allowance for $d-s$ mixing. If we regard $Na_4[Cu(NH_3)_4][Cu(S_2O_3)_2] \cdot H_2O$ as containing a square coplanar CuN_4 chromophore, the spectrum can be fitted with $e_\sigma = 6.4$ kK and $e_{ds} = 1.4$ kK. These may be compared with the values of 5.4 kK and 1.5 kK respectively which have been found for $CuCl_4^{2-}$ (vide infra – 2.5.2). Even in CuN_4X_2 chromophores, the large tetragonal distortion ensures that the d_{z^2} orbital is appreciably stabilized by $d-s$ mixing, thus raising the energy of the $^2A_{1g}$ state.

2.5.2 Copper(II) Halide Complexes

Chlorocuprates(II) exhibit an astonishing variety of structures (96). This behaviour may well be parallelled in bromocuprates(II), although less structural information is available. Fluorocuprates(II) always appear to contain axially-elongated CuF_6 chromophores (97).

Chlorocuprates(II) have been widely studied with a view to testing the transferability of crystal field or AOM parameters over a variety of structures. The earliest application of the AOM to chlorocuprates(II) involved the Ξ^2 model, with neglect of π-overlap (98). The $d-d$ transition energies of several species could be fitted with a single freely-chosen parameter σ^*, but much of the structural and spectroscopic data on which these calculations were based have now been superseded, and the Ξ^2 model does not work for some structural types which have been found more recently, such as the square coplanar $CuCl_4^{2-}$ ion. A more elaborate AOM treatment, including σ- and π-overlap as well as crystal field effects, was later applied to several chlorocuprates (99, 100); some anomalies in the earlier work were remedied, but the same objections apply in the light of more recent work. Similar remarks are applicable to the use of the point bond model (equivalent to the e_λ model) in the interpretation of the $d-d$ spectra of halocuprates(II) (43). In 1974, the crystal structure of $(C_6H_5CH_2CH_2NH_2Me)_2CuCl_4$ (101) revealed the first example of square coplanar $CuCl_4^{2-}$. This species has also been found in $(Et_2NH_2)_2CuCl_4$, although the low-temperature phase of this interesting compound also contains anions which are distorted towards tetrahedral geometry (102). Square-coplanar $CuCl_4^{2-}$ is characterised by a $d-d$ band at ca. 17 kK, higher in energy than any other $d-d$ absorption in chlorocuprates(II). Other $d-d$ bands are observed at around 12 kK and 14 kK (101, 103–105). There seems to be general agreement (34, 106, 107) that the relative energies of the excited states are $^2B_{2g} < {}^2E_g < {}^2A_{1g}$, implying that the d_{z^2} orbital lies lowest in energy. The transition energies cannot be satisfactorily fitted with two parameters e_σ and e_π, parameters which are consistent with the spectra of other chlorocuprates(II) predict that d_{z^2} should lie higher in energy than $d_{xz,yz}$. By introducing $3d-4s$ mixing, and the new AOM parameter e_{ds}, we obtain $e_\sigma = 5.4$ kK, $e_\pi = 0.9$ kK and $e_{ds} = 1.5$ kK (34). The fact that the ratio (e_π/e_σ) is lower than the value of 0.25 as calculated from the squared overlap integrals for σ- and π-overlap with ligand $3p$-orbitals reflects the contribution of ligand 3 s-overlap to e_σ. These parameters, *mutatis mutandis*, also fit the $d-d$ spectra of several other chlorocuprates(II) (34). In these calculations, the parameters $e_{dp\sigma}$ and $e_{dp\pi}$ were introduced to take account of $3d-4p$ mixing in noncentrosymmetric systems. An anomaly in previous AOM (and crystal field) calculations on chlorocuprates(II) had been the prediction that Δ_t, the splitting parameter for the hypothetical regularly tetrahedral $CuCl_4^{2-}$ ion, should be about 6 kK, which seems excessive compared with the values of 3–3.5 kK found for other MCl_4^{2-} ions. The introduction of $d-p$ mixing stabilizes the t_2 level relative to e, and leads to a much more reasonable value for Δ_t (see also 2.6). Thus it seems that in chlorocuprates(II), we need at least four parameters: $e_\sigma(S)$, $e_\sigma(p)$, e_π and e_{ds}, in lower symmetries, $e_{dp\sigma}$ and $e_{dp\pi}$ may also be required. The latter two parameters are relatively small, about 0.7 kK and 0.1 kK. We never have enough spectroscopic data for any one system to enable us to evaluate all these parameters; however, they should be transferable over a wide range of systems with different coordination geometries, making due allowance for changes in bond lengths.

Fluorocuprates(II) invariably contain axially-elongated CuF_6 chromophores; earlier reports of axially-compressed coordination geometries in K_2CuF_4 and

Ba_2CuF_6 have been refuted (97). However, in copper-doped $KZnF_3$, e.s.r. evidence shows that the Cu atoms have an axially-compressed coordination sphere in crystals containing up to about 30% copper (108). Although fluorocuprates(II) do not exhibit the rich variety of coordination geometries and coordination numbers which is found in chlorocuprates(II), the axial and equatorial bond lengths vary sufficiently to make profitable the use of the AOM to rationalise a large body of spectroscopic data with a few parameters. The spectra can be fitted rather well with only one freely-chosen parameter, estimating the relative values of $e_\sigma(eq)$ and $e_\sigma(ax)$, as well as the ratios e_π/e_σ, from the overlap integrals (109). The larger ratio (e_π/e_σ) compared with that required in chlorocuprates(II) may reflect the smaller extent of participation by fluorine $2s$-orbitals in the bonding (cf. 2.1.1).

2.5.3 Copper(II) Systems with Oxygen-Donor Ligands

Oxygen-donor ligands present problems in applying the AOM, since they are generally non-linear ligators. They are capable of functioning as π-donors, but it is usually necessary to define two e_π parameters. Even in mixed oxides, we cannot regard the O^{2-}-ions as linear ligators, since they are mostly bonded to other metal atoms of relatively high electronegativity. However, some very interesting applications of the AOM involve CuO_n chromophores.

   Mixed oxides containing Cu(II) have been thoroughly studied, and the AOM has proved useful in the analysis of their $d-d$ spectra (110–117). These usually contain axially-elongated CuO_6 octahedra, although square-coplanar, four-coordinate copper(II) occurs in $MCuSi_4O_{10}$ (M = Ca, Ba). In the tetragonal CuO_6 chromophores, the $d-d$ spectra can be analysed to yield $e_\sigma(eq)$ = ca. 6–7 kK, $e_\pi(\|)$ 2–2.5 kK and $e_\pi(\perp)$ = 3–4 kK. In practically all CuO_n chromophores, regardless of the nature of the O-donor ligand, $e_\pi(\perp)$ is greater than $e_\pi(\|)$. In the tetragonal mixed oxides, the d-orbital sequence appears to be $d_{x^2-y^2} > d_{z^2} > d_{xy} > d_{xz,yz}$. However, the energy sequence in the square coplanar CuO_4 chromophore in $MCuSi_4O_{10}$ is almost certainly $d_{x^2-y^2} > d_{xy} > d_{xz,yz} > d_{z^2}$; the $d-d$ spectrum (110, 115) can be fitted with e_σ = 9.4 kK, $e_\pi(\|)$ = 3.8 kK and $e_\pi(\perp)$ = 6.2 kK. The e_π parameters seem rather large, and it may be that $d-s$ mixing is important. If we regard the ligands as linear ligators (which is not strictly appropriate, since they are covalently bonded to silicon atoms in $MCuSi_4O_{10}$), we can fit the spectrum with e_σ = 6.2 kK, e_π = 1.5 kK and e_{ds} = 1.6 kK. These seem quite reasonable compared with the parameter values for square coplanar $[Cu(NH_3)_4]^{2+}$ and $CuCl_4^{2-}$ discussed above (2.5.1. and 2.5.2); e_{ds} seems to be remarkably constant. The higher value of e_π/e_σ for square coplanar CuO_4 compared with $CuCl_4^{2-}$ may reflect a smaller degree of participation by the ligand $2s$-orbitals in the former; we can separate e into the sum of $e_\sigma(p)$ and $e_\sigma(s)$, these taking the values 5.2 kK and 1.0 kK respectively, which seem not unreasonable.

   Reinen and *Grefer* (116) have measured the $d-d$ spectra of distorted tetrahedral CuO_4 chromophores in mixed oxides. This work concerns copper-doped lattices such as Zn_2SiO_4 and ZnO, where the exact coordination geometry about the copper is uncertain. However, an effort was made to relate the $d-d$ spectra to the distortion from T_d using the AOM. It was concluded that in copper-doped ZnO, trigonally-compressed CuO_4 chromophores are found.

To sum up the situation for Cu(II) in mixed oxides, it would seem that the application of the AOM has been quite successful, but it would be interesting to look in more detail at the possible effects of $d-s$ mixing.

Some difficulties have arison in the application of the AOM to copper(II) complexes with carboxylate ligands. Here we are undoubtedly dealing with non-linear ligators. $CaCu(CH_3COO)_4 \cdot 6H_2O$ and Cu(6-aminohexanoic acid)$_4(ClO_4)_2$ have similar structures, which may be regarded as distorted dodecahedra. The Cu atoms form short bonds to four carboxylate oxygens which form a highly-flattened tetrahedron, while four carboxylate oxygens at much greater distances (ca. 2.8 Å) form an elongated tetrahedron. The analysis of the $d-d$ spectrum of $CaCu(CH_3COO)_4 \cdot 6H_2O$ using the AOM lead to good agreement, including σ- and π-overlap and crystal field effects, using parameters consistent with the spectra of tetragonal CuO_6 chromophores (118, 119). However, it has been pointed out that the same parameters, *mutatis mutandis*, do not fit the spectrum of Cu(6-aminohexanoic acid)$_4(ClO_4)_2$ very well (120). The $d-d$ spectra of tetragonal CuO_6 chromophores can be satisfactorily rationalised within the AOM (113) but there are still problems concerning the e_π parameters for non-linear ligators. A further complication in the CuO_8 systems is the need to consider spin-orbit coupling (23), but this does not seem to be so important in the tetragonal systems.

Some interesting work has been done on the $[Cu(H_2O)_6]^{2+}$ ion (121–123). The three Cu–O distances in the rhombic cation are appreciably different, and good crystallographic data are available for five compounds. These offer an opportunity to test the idea that the AOM parameters vary with the internuclear distance in accordance with the square of the overlap integral, i.e. $e_\lambda = K_\lambda S_\lambda^2$. It was assumed (121) that in each Cu–OH$_2$ bond, the four atoms are coplanar, so that the parameters $K_{\pi s}$ and $K_{\pi c}$ can be defined, relating to π-interactions which are respectively parallel and perpendicular to the H–H vector in the water molecule. The polarized crystal spectra could be fitted to the three parameters K_σ, $K_{\pi s}$ and $K_{\pi c}$. Surprisingly, the latter two parameters were found to be nearly equal, and slightly greater than K_σ; all three were consistent with those required for other copper-oxygen chromophores. Another study (123) has led to substantially similar results; here, however, the O atoms were assumed to be tetrahedrally surrounded by two H atoms, the O–Cu bond, and the second lone pair. Both studies (121–123) made use of the AOM parameters in the interpretation of magnetic data (see also Sect. 5).

The electronic spectra of bis(acetylacetonato)-copper(II) and its derivatives have aroused controversy for about twenty years. In spite of much effort, polarized crystal spectra have not led to definitive assignments. *Hitchman* (124) has attempted to shed some light on the problem by comparing the crystal spectra of [Cu(acac)$_2$] and its quinoline adduct, using the AOM to determine how the addition of a quinoline molecule should affect the spectrum. He concluded that the energy sequence $d_{xy} > d_{z^2} >$

$d_{x^2-y^2} > d_{yz} \sim d_{xz}$ was most probable for [Cu(acac)$_2$]. Note that d_{z^2} appears to be the lowest-energy d-orbital in CuCl$_4^{2-}$ (D$_{4h}$) and in square-coplanar CuO$_4$ chromophores in mixed oxides. The neglect of $d-s$ mixing in Hitchman's analysis could be significant; without such mixing, it is difficult to place d_{z^2} lowest in energy without rather unreasonable values of the e_π parameters.

2.5.4 Five-Coordinate Copper(II) Systems

Apart from [Cu(acac)$_2$(quinoline)] which we have just mentioned, and trigonal bipyramidal CuCl$_5^{3-}$, to which the discussion in 2.5.2 is relevant, relatively little use has been made of the AOM in discussing the $d-d$ spectra of five-coordinate Cu(II) complexes. We have previously referred to Hitchman's work on [Cu(salen)(pyridine)] and [Cu(salen)]$_2$ in 2.3.2.

An interesting trigonal bipyramidal CuN$_5$ chromophore occurs in Ag[Cu(NH$_3$)$_2$-(NCS)$_3$], with the ammonia ligands in the axial positions. The polarized crystal spectrum (125) indicates the energy sequence $d_{z^2} > d_{xz,yz} > d_{xy,x^2-y^2}$. This is impossible to explain by simple AOM arguments, with thiocyanate assumed to be a linear ligator. $3d-4p$ mixing could conceivably lower d_{xy,x^2-y^2} below $d_{xz,yz}$ in energy, but $e_{dp\sigma}$ would have to be very large. It can be shown (126) that the introduction of symmetry-restricted covalence, which necessitates the inclusion of two-electron terms, can lead to the correct ordering. However, as with the tetragonal copper(II) ammine systems discussed in 2.5.1, the simple AOM fails.

The AOM has recently been used in the analysis of the polarized crystal spectrum of aquobis(1,10-phenanthroline)copper(II) nitrate (127), where the coordination geometry is trigonal bipyramidal. The water molecule occupies an equatorial position, so that the appropriate point group is C$_2$ although C$_{2v}$ is a reasonable approximation. However, it was found impossible to fit the spectrum with AOM parameters which could be regarded as chemically meaningful.

2.5.5 Miscellaneous Copper(II) Systems

The 'octahedral' [Cu(NO$_2$)$_6$]$^{4-}$ ion in compounds such as Cs$_2$PbCu(NO$_2$)$_6$ has aroused much interest. Although crystallographic studies indicated regular octahedral coordination about the copper, the $d-d$ spectrum is more consistent with a substantial tetragonal elongation; the e_g orbitals are apparently split by about 8 kK. The AOM has been used (128) to calculate the equatorial and axial Cu$-$N distances from the $d-d$ spectrum. The configuration of this ion has been a subject of continuing interest; the reader is referred to some recent work (129, 130) for the latest appraisal of the situation.

2.6 Other Aspects of d–d Spectra

Here we discuss two topics: the relationship between octahedral and tetrahedral ligand field splitting parameters, and the pressure-dependence of the octahedral parameter; a common feature of these is the use of the AOM to determine the variation of orbital splittings with the internuclear distance.

At an early stage in the development of ligand field theory, it was found that Δ_t, the splitting parameter for tetrahedral MX_4, should be equal to (4/9) of Δ_0, the splitting parameter for octahedral MX_6, and experimental data are in good agreement with this prediction. However, this takes no account of the fact that the M–X distance in tetrahedral MX_4 is usually some 8–10% shorter than in octahedral MX_6. In the pointcharge crystal field model, Δ is proportional to R^{-5}, so that if the difference in R between MX_4 and MX_6 is taken into account, we predict (Δ_t/Δ_0) to be 0.6–0.7, compared with the experimental value of about 0.5. An AOM treatment (*131*) leads to better results, since here we find:

$$\Delta_t = (4/3)e_\sigma - (16/9)\,e_\pi$$
$$\Delta_0 = 3\,e_\sigma \quad - \quad 4\,e_\pi$$

If we take the e_λ parameters to vary with the internuclear distance in accordance with the square of the appropriate overlap integral (assuming that only ligand *np*-orbitals are involved), we find that the π-overlap integrals always increase proportionately faster than the σ-overlap integrals as the internuclear distance is reduced, so that the negative contribution to Δ_t made by the e_π term becomes more important. For MCl_4 and MCl_6 chromophores (M = Mn(II), Co(II), Ni(II)), such calculations lead to values for (Δ_t/Δ_0) of about 0.57, compared with the experimental values of around 0.47 and the crystal-field value of about 0.70. The calculated ratio is still too high. This may be explained in terms of *d–p* mixing; if this is taken into account, we find:

$$\Delta_t = (4/3)\,e_\sigma - (16/9)\,e_\pi - (16/9)\,e_{dp\sigma} - (64/27)\,e_{dp\pi}$$

If we neglect the small $e_{dp\pi}$ term, we require $e_{dp\sigma}$ to be about 0.5 kK, which may be compared with the value of 0.75 kK found for $e_{dp\sigma}$ in D_{2d} $CuCl_4^{2-}$ (*34*).

The dependence of the octahedral splitting parameter Δ_0 on the internuclear distance R in a single compound can sometimes be determined by measuring the *d–d* spectrum and the crystallographic parameters over a range of applied pressures. *Drickamer* (*132*) noted that in NiO, Δ_0 was very nearly proportional to R^{-5}, as predicted by crystal field theory. However, the same result can be obtained from the AOM (*133*), considering only σ-overlap with oxygen $2p$-orbitals. This treatment may be compared with that of *Burns* and *Axe* (*134*), who used the Wolfsberg-Helmholz model to determine the theoretical variation of Δ_0 with R.

3. f–f Spectra

Some of the earliest applications of the AOM concerned the analysis of f–f splittings in compounds of the lanthanides and actinides, and interest in this area has been well-maintained. The approximation of weak covalent bonding would seem to be appropriate in dealing with f-orbitals; however, the same argument can be put forward in favour of the crystal field model, which continues to be widely used by workers in f–f spectra. There has therefore been much controversy as to whether the small f-orbital splittings (ca. 100–400 cm^{-1} in lanthanides, and 2000–10000 cm^{-1} in actinides) represent weak covalency effects or the results of a crystal field perturbation. However, there is a good deal of independent evidence for weak covalent bonding in compounds of the lanthanides; earlier work was reviewed by *Axe* and *Burns* (*135*), and more recent references will be found in a discussion of ligand hyperfine interactions with lanthanide ions (*136*). Moreover, the nephelauxetic effect is well-documented for $4f$- and $5f$-ions (*137*).

The Ξ^2 model appeared for the first time (*8*) in a discussion of orbital splittings in 8- and 9-coordinate lanthanide compounds. It was possible to fit the experimentally-determined orbital splittings quite well to only one parameter, σ^*, compared with the three parameters required by the crystal field model. The experimental values of σ^* thus found were in fair agreement with values estimated from the relationship between this formulation of the AOM and the Wolfsberg-Helmholz semi-empirical MO approach. However, the validity of this last statement is doubtful, especially following the results of photoelectron spectroscopy on lanthanide complexes which show, surprisingly, that antibonding orbitals can have larger ionization energies than their bonding counterparts − "the third revolution in ligand field theory" (*138*). The early Ξ^2 model considered only σ-bonding, but more general formulations of the AOM, including σ-, π-, δ- etc. overlap have been developed for f-orbital splittings (*20, 31, 32*). The Ξ^2 model was originally applied only to coordination geometries where no off-diagonal elements were required, i.e. where each irreducible representation occurs only once in the split f-orbitals. *Perkins* and *Crosby* (*139*) studied the f–f spectra of some ytterbium(III) chelates YbL_3, of D_3 symmetry, where off-diagonal elements arise. They adopted a contact perturbation model (essentially equivalent to the Ξ^2 model) and obtained poorer agreement with the experimental data than that afforded by the crystal field model, using Hartree-Fock wave functions to calculate the crystal field parameters rather than leaving them (as is customary) as freely-chosen parameters. Similar results were obtained from work on analogous terbium(III) chelates.

Cubic systems are, in principle, easier to deal with since there are only two splitting parameters for f-orbitals and no off-diagonal elements occur. At the same time, the problem is somewhat over-determined, since there are usually many spectroscopic transition energies from which to extract the AOM parameters. *Warren* (*32*) has recently applied the AOM, considering σ- and π-overlap, to some octahedral halide complex anions of lanthanides and actinides. For $PrCl_6^{3-}$, he finds $e_\sigma = 499$ cm^{-1} and $e_\pi = 262$ cm^{-1}, with (e_π/e_σ) equal to 0.525, a surprisingly high ratio. For actinide ions, the AOM parameters are larger, as expected, but display some curious trends. In PaX_6^{2-} (X = F, Cl, Br, I), the e_λ parameters decrease with increasing atomic number of the halogen, as in CrN_4X_2 chromophores (2.1.1). However, the ratio (e_π/e_σ) increases from 0.43 (F) to 0.85 (I), contrary to expectation. Similar behaviour was found for UX_6^- ions. In UX_6^{2-}, (e_π/e_σ) is 0.23 for UCl_6^{2-} and 0.24 for UBr_6^{2-}; the smaller ratios compared with UX_6^- are consistent with the lower oxidation state of the metal. However, these results have their surprising features. More spectroscopic data for actinide (and especially lanthanide) MX_6 chromophores would be desirable. It may be that l-mixing is important.

The Ξ^2 model has been applied (*141, 142*) to the f–f spectra of several trivalent lanthanide ions incorporated into the 8-coordinate cubic sites in MO_2 (M = Th, Ti, Zr, Ce). The values obtained for σ^* were in the range 60–$100\,cm^{-1}$ (i.e. $e_\sigma = 400$–$700\,cm^{-1}$), about twice as large as are found in enneaaquo complexes; this can be correlated with the larger nephelauxetic effect found in the mixed oxides.

We have noted that the number of orbital splitting parameters which can be extracted from an f–f spectrum (often with some difficulty and uncertainty) is relatively small; in low symmetries, there are more splitting parameters but the number of off-diagonal elements between f-orbitals of the same irreducible representation may be inconveniently large. The D_{2d} 8-coordination found for Ln in $LnXO_4$ (X = P, As, V) offers a good compromise between these considerations; the f-orbitals are split into five levels and there is only one off-diagonal term. *Kuse* and *Jørgensen* (*143*) used the AOM in a discussion of the spectrum of Er^{3+} in this environment; this work has been extended to Tm^{3+}, Eu^{3+} and Nd^{3+} (*144*). The results have been rather successful, and strengthen the feeling that the f-orbital splitting is best attributed to weak covalency rather than to crystal field effects. *Linares et al.* (*144*) used the equations of *Kibler* (*20, 30*) which relate the crystal field parameters to AOM parameters e_λ, and found that the relative magnitudes of the latter were in accordance with expectation. For example, with $Eu^{3+} : YPO_4$, they obtained $e_\sigma = 244\,cm^{-1}$, $e_\pi = 26\,cm^{-1}$, $e_\lambda = -19\,cm^{-1}$ and $e_\phi = 1\,cm^{-1}$. The relative magnitudes of the e_λ are satisfying. Variations in the magnitudes of the e_λ with the internuclear distance could also be readily rationalised. Thus the AOM parameters seem to have some chemical significance, which is more than can be said for the crystal field parameters.

Eu^{3+}-doped $LaCl_3$ (C_{3h} or D_{3h}), $LaAlO_3$ (D_3) and La_2O_3 (C_{3v}) have also been studied (*145*). In $LaCl_3$, there are no off-diagonal elements between f-orbitals (in the pseudo-symmetry D_{3h}); in the other compounds, two off-diagonal elements arise. The fitting of the experimental splitting parameters to σ^* (in the Ξ^2 formulation) was fairly successful for $LaCl_3$ and $LaAlO_3$, but not for La_2O_3.

4. Other Spectra and Optical Properties

4.1 Charge Transfer Spectra

Howald and *Keeton* (*43*) applied the 'point bond' model (essentially equivalent to the e_λ model) to the charge transfer spectra of copper(II) halide complexes. Three charge transfer bands were observed for most species, and were assigned to transitions to the highest (singly-occupied) metal d-orbital from (respectively) ligand nonbonding, π-bonding and σ-bonding orbitals. The transition energies were expressed as the sum of a term (constant for a given ligand) representing the energy gap between the donor orbital and the unperturbed metal d-orbitals, and a term representing the destabilization of the highest-energy d-orbital. The latter could be expressed in terms of AOM parameters whose magnitudes were obtained from the $d-d$ spectra. This procedure proved to be rather successful; there does appear to exist a correlation between the positions of the charge transfer bands and the highest $d-d$ transition energy.

A somewhat related treatment of the charge transfer spectra in some chloro-cuprates(II), using the Ξ^2 model, was performed by *Day* and *Jørgensen* (*98*).

4.2 Spectra of M−O−M Bridged Complexes

The electronic spectra of species such as $[(NH_3)_5 Cr-O-Cr(NH_3)_5]^{4+}$ cannot strictly be regarded as $d-d$ spectra since there is evidently much electron delocalisation over the M−O−M bridge. *Schmidtke* (*146*) and *Glerup* (*147*) have applied the AOM in the analysis of such spectra. The ground state of the abovementioned Cr(III) complex cation is a singlet, with a triplet excited state lying some $450\,\mathrm{cm}^{-1}$ higher in energy. The excited states can be treated in terms of antiferromagnetic coupling between the quartet and doublet states of the two $CrN_5 O$ chromophores. *Glerup* (*147*) showed that a one-electron matrix element between the d-orbitals on the two Cr atoms led to the Landé rule for antiferromagnetic coupling, and enables transitions to doubly-excited levels to become allowed through mixing with charge transfer states. Other work on this system, concentrating on magnetic rather than optical properties, is discussed in Section 5. *Schmidtke* (*146*) has compared the situation in linear-bridged systems with that in angled systems such as $[(NH_3)_5 Cr-OH-Cr(NH_3)_5]^{5+}$.

4.3 Vibronic Structure in Spin-Forbidden Transitions

Flint and *Matthews* (*148*) have used the AOM in a qualitative way in the analysis of the vibronic structure in the ${}^4A_{2g} \rightarrow {}^2E_g$ absorption and ${}^2E_g \rightarrow {}^4A_{2g}$ luminiscence of octahedral Cr(III) chromophores. It can be predicted that ligand vibrational modes (for example, the rocking mode of a coordinated NH_3 molecule) will appear strongly in the vibronic structure if the metal-ligand overlap integral fluctuates in magnitude as the vibration is executed. This is found to be the case for $[Cr(H_2O)_6]^{3+}$, $[Cr(D_2O)_6]^{3+}$ and $[Cr(en)_3]^{3+}$ and it is suggested that the analysis of the vibronic structure can be facilitated by such considerations.

4.4 Circular Dichroism in Chiral Chromophores

Schäffer (*149*) has discussed the circular dichroism of octahedral Cr(III) and Co(III) complexes of the types tris(bidentate) and *cis*-bis(bidentate). The chirality can be thought of as arising from displacements of the donor atoms from orthoaxial positions, and the effects of such displacements can be treated using the AOM.

4.5 p–p Spectra

A number of Se(II) and Te(II) complexes with dithiocarbamate (and related) ligands apparently form paramagnetic $MS_2(C_{2v})$ chromophores in solution (*150*). The electronic spectra of these systems can be treated as p–p spectra, and have been successfully interpreted using the AOM.

5. Magnetic and E.S.R. Properties

The AOM has been used to parameterise the ligand field in detailed analyses of magnetic properties in non-cubic complexes. It may also be used to obtain the ground state wave function in systems of low symmetry where d-orbital mixing is important; the mixing coefficients can be obtained from the off-diagonal elements of the AOM matrix in terms of e_λ parameters, whose magnitudes can be found from the d–d spectrum.

    Hitchman and his co-workers (*121, 122, 151*) have shown how the ground state wave function (in the form $ax^2 + by^2 + cz^2$) can be obtained for rhombic copper(II) systems; the coefficients a, b and c thus obtained are in reasonable agreement with those found by analysis of the e.s.r. spectrum. *Marshall* and *James* (*123*) have attempted an ambitious analysis of the optical and magnetic properties of $[Cu(H_2O)_6]^{2+}$ in several crystalline environments; the AOM was used to parameterise the theoretical expressions for the various experimental properties, in order to see whether a great deal of data could be fitted to a small number of parameters. The model was success-ful in this respect, but the AOM parameters seem to be rather unrealistic. For exam-ple, e_π was found to be more important than e_σ at long internuclear distances, which seems unlikely.

Gerloch and his co-workers (*152–154*) have developed an elaborate framework for the analysis of the magnetic properties of low-symmetry complexes, using the AOM to parameterise the ligand field. The model has been applied to $M(py)_4 X_2$ (M = Fe, Co and X = Cl, Br, NCS), which have very little symmetry at all. Pyridine turns out to have a positive value for e_π, so that it is apparently a π-donor towards Fe(II) and Co(II). This is in agreement with optical studies of $Ni(py)_4 X_2$ (*48, 66*) but in dis-agreement with the work of *Glerup et al.* (*58*) on $[Cr(py)_4 X_2]^+$ where pyridine is apparently a π-acceptor. It was found that e_π for Br^- was greater than that for Cl^-, which does not appear to be the case in Ni(II) or Cr(III) systems. Rather surprisingly, the best fits to the experimental data were found with the systems of lowest symme-try; the magnetic properties of these systems are very sensitive to small deviations from idealised higher symmetries.

Horrocks (*155, 156*) has developed a somewhat similar comprehensive approach to the optical and magnetic properties of low-symmetry complexes, using a weak-field model. This has been applied to the $CoCl_4^{2-}$ ion in $Cs_2 CoCl_4$ and $Cs_3 CoCl_5$. The ligand field was parameterised in terms of e_σ only, and values of this parameter, together with the spin-orbit coupling constant λ, the orbital reduction factor k and the Racah parameter B were obtained by fitting the $d–d$ spectra, zero-field splittings, principal magnetic susceptibilities and e.s.r. g-values.

Kahn and *Briat* (*157, 158*) have applied the AOM in a discussion of the antiferro-magnetism of binuclear complexes such as $Cr_2 OL_{10}, [Cr_2 X_9]^{3-}$ and $Fe_2 OL_{10}$. In contrast to *Glerup* (*147*), *Kahn* and *Briat* suggest that the exchange interactions in $[(NH_3)_5 Cr–O–Cr(NH_3)_5]^{4+}$ can be interpreted by considering only the ground state configuration, and mixing with charge transfer states is only a small second-order ef-fect.

Meriaudeau et al. (*159*) studied the e.s.r. spectrum of Ti(III) in anatase, where the coordination geometry is a distorted octahedron of approximately D_{2d} symmetry. The AOM was used (in the Ξ^2 formulation) to parameterise the $d–d$ transition ener-gies which appear in the expressions for the g-values. It was found that the deviations from $g = 2.00$ were smaller than expected, although this might be attributed to the neglect of π-bonding in the Ξ^2 model. *Kibler* and *Kodratoff* (*160*) obtained better calculated g-values by using a different value of the crystallographic parameter which describes the displacements of the four equatorial O atoms out of the plane to give D_{2d} symmetry.

6. Miscellaneous Applications

6.1 Stereochemistry

Molecular orbital theory has been used to discuss the relative energies associated with alternative geometries by looking at the variation of bonding and antibonding orbital energies as functions of the structural parameters. The AOM has proved useful in parameterising orbital energies for this purpose.

Kettle and his co-workers (*39–42*) used a model rather similar to the AOM to discuss stereochemistry. A perturbation approach led to the proportionality of MO energies (relative to the unperturbed orbitals) to squared overlap integrals, as in the AOM. For systems where the valence shell orbitals are evenly occupied, the total stabilization energy shows no angular dependence, suggesting that steric forces determine the equilibrium geometry.

Burdett (*35–38*) has applied the AOM in a similar way, but to a wider variety of d^n configurations. He shows that the total electronic stabilization energy arising from σ-overlap in a transition metal complex is given by:

$$\sum(\sigma) = \beta_0 \sum_j h_j S_j^2 [d_j(\Gamma_j); \ \sigma(\Gamma_j)]$$

where S is the overlap integral between the metal orbital $d(\Gamma_j)$ belonging to the irreducible representation Γ_j and the ligand group orbital $\sigma(\Gamma_j)$, while h_j is the number of holes in the metal orbital $d_j(\Gamma_j)$. Thus bonding is maximised when the squared overlap between ligand orbitals and holes is maximised. For species where $\sum(\sigma)$ is independent of the angular geometry, the inclusion of a quartic term in S^4 helps to decide between two plausible geometries. The model is rather successful in predicting the most stable geometries for systems of a given coordination number and a given number of d-electrons. It can also be used to determine the relative stabilities of isomers, such as *cis-* and *trans*-MA_4B_2, and to rationalise some features of intramolecular rearrangement reactions involving interconversion of such isomers. Burdett has criticised the use of the Jahn-Teller theorem in the rationalisation of stereochemistry for, e.g., d^9 systems; the present author expressed similar feelings in an earlier discussion of overlap considerations which might be relevant to such structural distortions (*161*). Burdett has shown that the total stabilization energies for five- and six-coordinate d^8 and d^9 systems are the same as those for four-coordinate square coplanar systems, suggesting that the fifth and sixth ligands are weakly bound; this may be significant in view of the characteristic tetragonal elongation in six-coordinate copper(II) systems, and the well-known equilibria between four- and six-coordinate species in Ni(II) chemistry.

Burdett has also applied the AOM to the rationalisation of the geometries of main group molecules (*38*). Correlation diagrams (analogous to Walsh diagrams) can be constructed to show the angular dependence of MO energies, and the shapes of simple molecules can be correctly predicted. There are complications arising from involve-ment of central-atom s-orbitals, but the model offers an appealing alternative to other theories of molecular geometry.

6.2 Thermodynamic Properties

The concept of crystal field stabilization energy has long been used in the rationalisa-tion of trends in the hydration enthalpies of transition metal ions, the lattice energies of simple salts, and the preferences of transition metal ions for octahedral or tetrahed-ral sites in complex crystals. *Burdett (162)* has used the AOM to calculate the stabili-zation energies of octahedral and tetrahedral complexes as a function of the number of d-electrons present. If only σ-bonding is considered, the stabilization energy for an octahedral complex (high-spin) remains constant from d^0 to d^3, decreases from d^3 to d^5, remains constant between d^5 and d^8, and falls again from d^8 to d^{10}. This is con-sistent with the known variation of the hydration enthalpies of divalent ions across the first transition series, superimposed upon a steadily increasing stabilization ener-gy. *Burdett* proposes that the difference between the experimental hydration enthal-py and the AOM stabilization energy reflects the interactions between ligand orbitals and the metal 4 s- and 4 p-orbitals, increasing as we go across the series. The model can also be applied to the interpretation of the reaction rates of aquo complexes and the preferences for octahedral vs. tetrahedral coordination in solids such as spinels. However, *Burdett* emphasises that the roles of 4 s- and 4 p-orbitals are important, and crystal field stabilization energies (whether discussed from an electrostatic or covalent viewpoint) which consider only the interactions of the metal d-orbitals may be misleading.

7. Concluding Remarks

Over half of this article has been devoted to the interpretation of d–d spectra; this is a fair reflection on the dominance of this area in applications of the AOM. However, we hope that readers who may have associated the AOM exclusively with electronic spectra will appreciate its value in other directions. Indeed, the AOM is applicable to all problems which depend on the relative energies of orbitals.

We now wish to look at some of the assumptions which are commonly made in using the AOM, and to discuss their validity. A common approximation for linear ligators is to estimate the ratio (e_π/e_σ) from the squared overlap integral ratio, $(S_\pi/S_\sigma)^2$, considering only overlap with ligand np-orbitals. It now seems that this ap-

112

proximation is invalid. In $CuCl_4^{2-}$ (D_{4h}), for example, (e_π/e_σ) is required to be 0.16, compared with the calculated value of 0.26 (34), while in $[Cr(NH_3)_4Cl_2]^+$ (58) this ratio is found to be (for the chloride ligands) about 0.2, compared with a calculated value of 0.35. Indeed, the magnitude of (e_π/e_σ) required to fit the spectra is always less than that calculated from the overlap integrals. As we have mentioned above (2.1.1), this can be explained if we suppose that ligand ns-orbitals are also involved in σ-bonding. If we wish to gauge the radial dependence of e_σ, this parameter should be separated into its s- and p-components, which should be proportional to $S_\sigma^2(s)$ and $S_\sigma^2(p)$ respectively.

Our work on chlorocuprates(II) (34) shows the need to include $d-s$ and $d-p$ mixing, with the introduction of the additional parameters e_{ds}, $e_{dp\sigma}$ and $e_{dp\pi}$. The fact that the AOM fails for square-coplanar $CuCl_4^{2-}$ if such mixing is neglected is rather disturbing. The need to consider explicitly the metal $(n + 1)$ s- and $(n + 1)$ p-orbitals strikes at the heart of one of the fundamental ideas of ligand field theory, the notion that properties of transition metal compounds can be interpreted by focussing our attention on the orbitals of the partly-filled shell, to the exclusion of all others. The paper in which the term 'angular overlap model' first appeared (13) was subtitled 'An Attempt to Revive the Ligand Field Approaches'. To appreciate the significance of this apt comment, we should remember the status of ligand field theory in the mid-1960's. Between 1955 and 1960, the simple electrostatic model was largely discarded, although its terminology and basic philosophy survived to some extent in the 'ligand field theory', 'adjusted crystal field theory' or 'expanded radial function model'. In the early 1960's, semi-empirical molecular orbital theories aroused great enthusiasm, and it was widely believed that the Wolfsberg-Helmholz model, or something like it, might supersede ligand field theory. Although the MO approaches considered all metal and ligand valence orbitals (in contrast to the ligand field theory, which concentrates on the metal d-orbitals), it was felt that the additional complexity of the model would be recompensed by a wealth of information. These hopes were not fulfilled, and by 1970 interest in the Wolfsberg-Helmholz and related models had waned. One reason for this was the preoccupation of theoretical inorganic chemists with the need to reproduce spectroscopic transition energies, and it soon became clear that the Wolfsberg-Helmholz model was inappropriate for this purpose. More recently, there has been a marked revival of interest in semi-empirical MO calculations on transition metal complexes (163); the difficulties associated with their use in calculating spectroscopic transition energies may have obscured their value in other directions. But in 1965, enthusiasm for such semi-empirical models was approaching its zenith, and the introduction of the AOM did, in a real sense, represent a revival of ligand field theory; once again we could focus our attention on the metal d-orbitals (or f-orbitals), although it was now necessary to inspect the ligand atom orbitals with care. Thus it may be felt in some quarters that the introduction of metal s- and p-orbitals is a retrograde step. There are now too many parameters for the interpretation of the spectrum of any one system. However, a useful analysis can be performed over a range of compounds which offer different structures, different bond lengths and a substantial number of spectroscopic parameters, as is found in the chlorocuprates(II) (*34*). We need consider d–s mixing only in cases where there is a large tetragonal component in the ligand field; such mixing may not be very important in most tetragonal complexes of Cr(III) and Ni(II), for example. d–p mixing is a smaller effect, and can probably be neglected for most purposes, except in tetrahedral systems. Thus the situation is perhaps not as bad as it may seem.

The assumption that AOM parameters are transferable from one system to another is one which deserves close scrutiny. For example, the negative value of e_π for pyridine found in $[Cr(py)_4X_2]^+$ depends on the assumption that the e_λ parameters for the halides are the same as in $[Cr(NH_3)_4X_2]^+$ (*58*). However, there is some reason to believe that the AOM parameters for halides may differ considerably from one compound to another in $Cr(III)N_4X_2$ chromophores, even when the equatorial ligands are closely related. Thus *Barton* and *Slade* (*57*) found e_σ for F^- to range between 7.2 kK and 9.0 kK, and e_π between 1.7 kK and 2.5 kK, in a series of CrN_4X_2 systems where the equatorial ligands were saturated amines. This could account for the negative value of e_π obtained for pyridine, which seems inconsistent with the positive values found for Fe(II), Co(II) and Ni(II) complexes (*48, 66, 153, 154*).

The future of the AOM would seem to lie along two directions:

(a) The analysis of copious amounts of experimental spectroscopic data over a series of closely related compounds of relatively high symmetry, to test the assumption of parameter transferability.

(b) The use of AOM to parameterise ligand fields of very low symmetry, along the lines suggested by *Gerloch* (*152–154*) and *Horrocks* (*155–156*). Such studies promise to be valuable in bio-inorganic chemistry, where transition metal ions are much used to probe the active sites of enzymes, usually in situations where there is little or no symmetry.

Ligand field studies of non-cubic complexes have experienced a marked resurgence in recent years; the use of more sophisticated group theoretical techniques (*164*), together with modern computing methods, should make it possible to perform analyses which might have daunted most workers a few years ago. The AOM was developed at a time when ligand field theory was in the doldrums, threatened by the apparently remorseless advance of molecular orbital theory and abandoned by coordination chemists whose interest had shifted towards organometallic chemistry and 'non-classical' complexes with ligands such as CO and PR_3; the AOM seems destined to play a significant part in what promises to be a fresh renaissance of ligand field theory.

References

1. *Jørgensen, C.K.:* Modern Aspects of Ligand Field Theory. Amsterdam: North Holland 1971
2. *Lever, A.B.P.:* Inorganic Electronic Spectroscopy. Amsterdam: Elsevier 1968
3. *Gerloch, M., Slade, R.C.:* Ligand Field Parameters. Cambridge University Press 1973
4. *Purcell, K.F., Kotz, J.C.:* Inorganic Chemistry p. 543. Philadelphia: W.B. Saunders 1977
5. *Yamatera, H.:* Naturwiss. **44**, 375 (1957)
6. *Yamatera, H.:* Bull. Chem. Soc. Japan **31**, 95 (1958)
7. *McClure, D.S.:* Advances in the Chemistry of Coordination Compounds (Kirschner, S., ed.) p. 498. New York: McMillan 1961
8. *Jørgensen, C.K., Pappalardo, R., Schmidtke, H.-H.:* J. Chem. Phys. **39**, 1422 (1963)
9. *Jørgensen, C.K., Schmidtke, H.-H.:* Z. Phys. Chem. (Frankfurt) **38**, 118 (1963)
10. *Schmidtke, H.-H.:* Z. Naturforsch. **19a**, 1502 (1964)
11. *Jørgensen, C.K.:* J. Phys. (Paris) **26**, 825 (1965)
12. *Schäffer, C.E., Jørgensen, C.K.:* Mat. Fys. Medd. Dan. Vid. Selsk. **34**, no. 13 (1965)
13. *Schäffer, C.E., Jørgensen, C.K.:* Mol. Phys. **9**, 401 (1965)
14. *Jørgensen, C.K.:* Structure and Bonding **1**, 3 (1966)
15. *Schäffer, C.E.:* Theoret. Chim. Acta **4**, 166 (1966)
16. *Jørgensen, C.K.:* Chem. Phys. Letters **1**, 1 (1967)
17. *Schäffer, C.E.:* Structure and Bonding **5**, 68 (1968)
18. *Lacroix, R.:* Comptes Rend. Acad. Sci. Paris **266**, 291 (1968)
19. *Schäffer, C.E.:* Pure Appl. Chem. **24**, 361 (1971)
20. *Kibler, M.R.:* Chem. Phys. Letters **8**, 142 (1971)
21. *Kibler, M.R.:* J. Chem. Phys. **55**, 1989 (1971)
22. *Schäffer, C.E.:* Internat. J. Quantum Chem. **5**, 379 (1971)
23. *Smith, D.W.:* Structure and Bonding **12**, 49 (1972)
24. *Harnung, S., Schäffer, C.E.:* Structure and Bonding **12**, 257 (1972)
25. *Schäffer, C.E.:* Structure and Bonding **14**, 69 (1973)
26. *Schäffer, C.E.:* Wave Mechanics – the First Fifty Years (Price, W.C., Chissick, S.S., Ravensdale, T., eds.) Ch. 12. London: Butterworths 1973
27. *Schäffer, C.E.:* Theoret. Chim. Acta **34**, 237 (1974)
28. *Kibler, M.R.:* J. Chem. Phys. **61**, 3859 (1974)
29. *Larsen, E., La Mar, G.N.:* J. Chem. Educ. **51**, 633 (1974)
30. *Kibler, M.R.:* Internat. J. Quantum Chem. **9**, 403 (1975)
31. *Urland, W.:* Chem. Phys. **14**, 393 (1976)
32. *Warren, K.D.:* Inorg. Chem. **16**, 2008 (1977)
33. *Smith, D.W.:* J. Chem. Soc. (A) 1509 (1969)
34. *Smith, D.W.:* Inorg. Chim. Acta **22**, 107 (1977)
35. *Burdett, J.K.:* Inorg. Chem. **14**, 375 (1975)
36. *Burdett, J.K.:* Inorg. Chem. **14**, 931 (1975)
37. *Burdett, J.K.:* Inorg. Chem. **15**, 212 (1976)
38. *Burdett, J.K.:* Structure and Bonding **31**, 67 (1976)
39. *Kettle, S.F.A.:* J. Chem. Soc. (A) 420 (1966)
40. *Kettle, S.F.A.:* J. Chem. Soc. (A) 1307 (1966)
41. *Kettle, S.F.A., Pioli, A.J.P.:* J. Chem. Soc. (A) 122 (1968)
42. *Kettle, S.F.A.:* Coord. Chem. Rev. **2**, 9 (1967)
43. *Howald, R.A., Keeton, D.P.:* Spectrochim. Acta **22**, 1211 (1966)
44. *Jørgensen, C.K.:* Prog. Inorg. Chem. **4**, 73 (1962)
45. *König, E.:* Structure and Bonding **9**, 175 (1971)
46. *Baker, W.A., Phillips, M.G.:* Inorg. Chem. **5**, 1042 (1966)
47. *Glerup, J., Schäffer, C.E.:* Progress in Coordination Chemistry (Cais, M., ed.) p. 100. Amsterdam: Elsevier 1968

48. *Lever, A.B.P.:* Coord. Chem. Rev. 3, 119 (1968)
49. *Perumareddi, J.R.:* Coord. Chem. Rev. 4, 73 (1969)
50. *Dubicki, L., Martin, R.L.:* Aust. J. Chem. 22, 839 (1969)
51. *Bunel, S., Ibarra, C., Adan, L.:* J. Inorg. Nucl. Chem. 31, 3203 (1969)
52. *Dubicki, L., Hitchman, M.A., Day, P.:* Inorg. Chem. 9, 118 (1970)
53. *Dubicki, L., Day, P.:* Inorg. Chem. 10, 2043 (1971)
54. *Keeton, M., Fa-Chun Chou, B., Lever, A.B.P.:* Canad. J. Chem. 49, 192 (1971); erratum, ibid. 51, 3690 (1973)
55. *Rowley, D.A.:* Inorg. Chem. 10, 397 (1971)
56. *Klein, R.L., Miller, N.C., Perumareddi, J.R.:* Inorg. Chim. Acta 7, 685 (1973)
57. *Barton, T.J., Slade, R.C.:* J. Chem. Soc. Dalton Trans. 650 (1975)
58. *Glerup, J., Monsted, O., Schäffer, C.E.:* Inorg Chem. 15, 1399 (1976)
59. *Soules, T.F., Richardson, J.W., Vaught, D.M.:* Phys. Rev. B3, 2186 (1971)
60. *Clementi, E.:* Tables of Atomic Functions, Supplement to I.B.M. J. Res. Development 9, 2 (1965)
61. *Wentworth, R.A.D., Piper, T.S.:* Inorg. Chem. 4, 1524 (1965)
62. *Schmidtke, H.-H.:* Chem. Phys. Letters 4, 451 (1969)
63. *Bertini, I., Gatteschi, D., Scozzofava, A.:* Inorg. Chem. 15, 203 (1976)
64. *Rowley, D.A., Drago, R.S.:* Inorg. Chem. 6, 1092 (1967)
65. *Rowley, D.A., Drago, R.S.:* Inorg. Chem. 7, 795 (1968)
66. *Hitchman, M.A.:* Inorg. Chem. 11, 2387 (1972)
67. *Schreiner, A.F., Hamm, D.J.:* Inorg. Chem. 12, 2037 (1973)
68. *Hamm, D.J., Bordner, J., Schreiner, A.F.:* Inorg. Chim. Acta 7, 637 (1973)
69. *Goodgame, D.M.L., Goodgame, M., Hitchman, M.A., Weeks, M.J.:* Inorg. Chem. 5, 635 (1966)
70. *Reiff, W.M., Long, G.J., Little, B.F.:* Inorg. Nucl. Chem. Letters 12, 405 (1976)
71. *Griffith, I.M., Nicholls, D., Seddon, K.R.:* J. Chem. Soc. (A) 2513 (1971)
72. *Russell, C.W.G., Smith, D.W.:* Inorg. Chim. Acta 6, 677 (1972)
73. *Seddon, K.R.:* Inorg. Chim. Acta 9, 123 (1974)
74. *Ziegler, M.L., Nuber, B., Wiedenhammer, K., Hoch, G.:* Z. Naturforsch. 32b, 18 (1977)
75. *Ciampolini, M.:* Structure and Bonding 6, 52 (1969)
76. *Norgett, M.J., Venanzi, L.M.:* Inorg. Chim. Acta 2, 107 (1968)
77. *Bertini, I., Dapporto, P., Gatteschi, D., Scozzofava, A.:* Inorg. Chem. 14, 1639 (1975)
78. *Bertini, I., Gatteschi, D., Scozzofava, A.:* Inorg. Chem. 14, 813 (1975)
79. *Bencini, A., Gatteschi, D.:* J. Phys. Chem. 80, 2126 (1976)
80. *Hitchman, M.A.:* Inorg. Chem. 16, 1985 (1977)
81. *Hougen, J.T., Leroi, G.E., James, T.C.:* J. Chem. Phys. 34, 1670 (1961)
82. *DeKock, C.W., Gruen, D.M.:* J. Chem. Phys. 44, 4383 (1966)
83. *DeKock, C.W., Gruen, D.M.:* J. Chem. Phys. 46, 1096 (1967)
84. *Jørgensen, C.K.:* Mol. Phys. 7, 417 (1964)
85. *Smith, D.W.:* Chem. Phys. Letters 6, 83 (1970
86. *Smith, D.W.:* Inorg. Chim. Acta 5, 231 (1971)
87. *Lever, A.B.P., Hollebone, B.R.:* Inorg. Chem. 11, 2183 (1972)
88. *Hargittai, I., Tremmel, J.:* Coord. Chem. Rev. 18, 257 (1976)
89. *Hathaway, B.J.:* Structure and Bonding 14, 49 (1973)
90. *Smith, D.W.:* Inorg. Chem. 5, 2236 (1966)
91. *Tomlinson, A.A.G., Hathaway, B.J.:* Coord. Chem. Rev. 5, 1 (1970)
92. *Hathaway, B.J., Billing, D.E.:* Coord. Chem. Rev. 5, 143 (1970)
93. *Hathaway, B.J., Stephens, F.S.:* J. Chem. Soc. (A) 884 (1970)
94. *Smith, D.W.:* Chem. Phys. Letters 16, 426 (1972)
95. *Smith, D.W.:* J. Chem. Soc. Dalton Trans. 1853 (1973)
96. *Smith, D.W.:* Coord. Chem. Rev. 21, 93 (1976)
97. *Friebel, C., Reinen, D.:* Z. anorg. allgem. Chem. 407, 193 (1974)

116

98. *Day, P., Jørgensen, C.K.:* J. Chem. Soc. 6226 (1964)
99. *Smith, D.W.:* J. Chem. Soc. (A) 2529 (1969)
100. *Smith, D.W.:* J. Chem. Soc. (A) 2900 (1970)
101. *Harlow, R.L., Wells, W.J., Watt, G.W., Simonsen, S.H.:* Inorg. Chem. **13**, 2106 (1974)
102. *Harlow, R.L., Simonsen, S.H.:* Abstract INOR 81, 173rd. Amer. Chem. Soc. Meeting 1977
103. *Hill, D.R., Smith, D.W.:* J. Inorg. Nucl. Chem. **36**, 466 (1974)
104. *Willett, R.D., Haugen, J.A., Lesback, J., Morrey, J.:* Inorg. Chem. **13**, 2510 (1974)
105. *Cassidy, P., Hitchman, M.A.:* Chem. Comm. 813 (1975)
106. *Watt, G.W., Wells, W.J.:* J. Inorg. Nucl. Chem. **38**, 921 (1976)
107. *Hitchman, M.A.:* private communication
108. *Reinen, D., Friebel, C., Propach, V.:* Z. Naturforsch. **31b**, 1574 (1976)
109. *Oelkrug, D.:* Structure and Bonding **9**, 1 (1971)
110. *Clark, M.G., Burns, R.G.:* J. Chem. Soc. (A) 1034 (1967)
111. *Reinen, D.:* Z. Naturforsch. **23a**, 521 (1968)
112. *Friebel, C., Reinen, D.:* Z. Naturforsch. **24a**, 1518 (1969)
113. *Smith, D.W.:* J. Chem. Soc. (A) 176 (1970)
114. *Reinen, D.:* Angew. Chem. Internat. Edn. **10**, 901 (1971)
115. *Grefer, J., Reinen, D.:* Z. Naturforsch. **28a**, 464 (1973)
116. *Reinen, D., Grefer, J.:* Z. Naturforsch. **28a**, 1186 (1973)
117. *Reinen, D., Weitzel, H.:* Z. anorg. allgem. Chem. **424**, 31 (1976)
118. *Smith, D.W.:* J. Chem. Soc. (A) 1024 (1971)
119. *Smith, D.W.:* J. Chem. Soc. (A) 1209 (1971)
120. *Garner, C.D., Lambert, P., Mabbs, F.E., Porter, J.K.:* J. Chem. Soc. Dalton Trans. 320 (1972)
121. *Hitchman, M.A., Waite, T.D.:* Inorg. Chem. **15**, 2150 (1976)
122. *Waite, T.D., Hitchman, M.A.:* Inorg. Chem. **15**, 2155 (1976)
123. *Marshall, R.C., James, D.W.:* J. Phys. Chem. **78**, 1235 (1974)
124. *Hitchman, M.A.:* Inorg. Chem. **13**, 2218 (1974)
125. *Hathaway, B.J., Billing, D.E., Dudley, R.J., Fereday, R.J., Tomlinson, A.A.G.:* J. Chem. Soc. (A) 806 (1970)
126. *McNeill, P.M., Smith, D.W.:* unpublished work
127. *Bencini, A., Gatteschi, D.:* Inorg. Chem. **16**, 1994 (1977)
128. *Smith, D.W.:* Theoret. Chim. Acta **17**, 89 (1970)
129. *Reinen, D., Friebel, C.:* Structure and Bonding, in press
130. *Mullen, D., Heger, G., Reinen, D.:* Solid State Commun. **17**, 1249 (1975)
131. *Smith, D.W.:* J. Inorg. Nucl. Chem. **34**, 3930 (1972)
132. *Drickamer, H.G.:* J. Chem. Phys. **47**, 1880 (1967)
133. *Smith, D.W.:* J. Chem. Phys. **50**, 2784 (1969)
134. *Burns, G., Axe, J.D.:* J. Chem. Phys. **45**, 4362 (1966)
135. *Axe, J.D., Burns, G.:* Phys. Rev. **152**, 331 (1966)
136. *McGarvey, B.R.:* J. Chem. Phys. **65**, 955, 962 (1976)
137. *Caro, P., Beaury, O., Antic, E.:* J. Phys. (Paris) **37**, 671 (1976) and references therein
138. *Jørgensen, C.K.:* Structure and Bonding **22**, 49 (1975)
139. *Perkins, W.G., Crosby, G.A.:* J. Chem. Phys. **42**, 407 (1965)
140. *Perkins, W.G., Crosby, G.A.:* J. Chem. Phys. **42**, 2621 (1965)
141. *Jørgensen, C.K., Pappalardo, R., Ritterhaus, E.:* Z. Naturforsch. **19a**, 424 (1964)
142. *Jørgensen, C.K., Pappalardo, R., Ritterhaus, E.:* Z. Naturforsch. **20a**, 54 (1965)
143. *Kuse, D., Jørgensen, C.K.:* Chem. Phys. Letters **1**, 314 (1967)
144. *Linares, C., Louat, A., Blanchard, M.:* Structure and Bonding **33**, 214 (1977)
145. *Linares, C., Louat, A.:* J. Phys. (Paris) **36**, 717 (1975)
146. *Schmidtke, H.-H.:* Theoret. Chim. Acta **20**, 92 (1971)
147. *Glerup, J.:* Acta Chem. Scand. **26**, 3775 (1972)
148. *Flint, C.D., Matthews, A.P.:* J. Chem. Soc. Farad. Trans. II **72**, 579 (1976)

149. *Schäffer, C.E.:* Proc. Roy. Soc. **A297**, 96 (1967)
150. *Nikolov, G.S., Smith, D.W.:* J. Chem. Soc. (A) 3250 (1971)
151. *Dawson, R., Hitchman, M.A., Prout, C.K., Rossotti, F.J.C.:* J. Chem. Soc. Dalton Trans. 1509 (1972)
152. *Gerloch, M., McMeeking, R.F.:* J. Chem. Soc. Dalton Trans. 2443 (1975)
153. *Gerloch, M., McMeeking, R.F., White, M.:* J. Chem. Soc. Dalton Trans. 2453 (1975)
154. *Gerloch, M., McMeeking, R.F., White, M.:* J. Chem. Soc. Dalton Trans. 655 (1976)
155. *Horrocks, W. DeW.:* Inorg. Chem. **13**, 2775 (1974)
156. *Horrocks, W. DeW., Burlone, D.A.:* J. Amer. Chem. Soc. **98**, 6512 (1976)
157. *Kahn, O., Briat, B.:* J. Chem. Soc. Farad. Trans. II **72**, 268 (1976)
158. *Kahn, O., Briat, B.:* J. Chem. Soc. Farad. Trans. II **72**, 1441 (1976)
159. *Meriaudeau, P., Che, M., Jørgensen, C.K.:* Chem. Phys. Letters **5**, 131 (1970)
160. *Kibler, M., Kodratoff, Y.:* J. Chim. Phys. Physicochim. Biol. **69**, 905 (1972)
161. *Smith, D.W.:* J. Chem. Soc. (A) 1498 (1970)
162. *Burdett, J.K.:* J. Chem. Soc. Dalton Trans. 1725 (1976)
163. *Smith, D.W.:* J. Chem. Soc. Dalton Trans. 834 (1976) and references therein
164. *Donini, J.C., Hollebone, B.R., Lever, A.B.P.:* Prog. Inorg. Chem. **22**, 225 (1977) and references therein

He(I) Photoelectron Spectra of d-Metal Compounds

Claudio Furlani and Carla Cauletti[1])

Institute of General and Inorganic Chemistry, University of Rome,
00185 Rome, Italy

Table of Contents

He(I) photoelectron spectra of d-metal compounds yield new, and in several instances unique experimental evidence on the electronic structure of the valence shell. UV P. E. spectroscopy (UPS) is therefore a technique of fundamental and general significance in the investigation of transition metal complexes, particularly if He(II) spectra are also considered, as a diagnostic tool for the identification of d-orbital levels, and despite the limitation to gas-phase measurements, hence to volatile compounds. Attention is called on some basic aspects of the interpretation of P. E. spectra of metal complexes, particularly the imperfect correspondence between orbital energies and ionization energies (deviations from Koopmans' behaviour), and the actual identification of d-orbital ionizations. Although the field is new and still in rapid development, considerable new light is thrown by the results of UPS investigations carried out from about 1970 till now on the electronic structures of several categories of coordination compounds, including particularly carbonyls, sandwich complexes and other compounds of transition metals in low oxidation state, and, among complexes of metals in higher or „normal" oxidation state, oxo- and halo-species, as well as several categories of inner-complex chelates.

---

[1]) CNR Laboratory of Theory and Electronic Structure of Coordination Compounds, Rome.

Knowledge of the electronic structure of d-metal compounds is based experimentally on a number of investigation techniques, among which electronic absorption spectroscopy has been the most widely used and the most effective in yielding information on the highest occupied and the lowest unoccupied energy levels of coordination compounds. Electron absorption spectroscopy focuses our attention on the partly filled shell levels of predominant metal d character and is instrumental in monitoring and developing bond models such as the ligand-field theory. The latest addition to the techniques which aid or supplement electron absorption spectroscopy is photoelectron (P.E.) spectroscopy, which gives information about filled valence orbitals in d-metal compounds that both confirms and extends the picture given by electronic absorption spectrscopy. P.E. spectroscopy also gives a more complete and immediate description of the whole valence shell of such compounds, including both d-levels and delocalized or ligand-localized valence orbitals and can therefore become even superior to the former technique as a source of information and as a tool for checking and promoting developments in bond theories. Although application of P.E. spectroscopy to d-metal compounds is still in its initial stage, a considerable amount of experimental data has been produced; some problems are now better focused, and promising results fo general significance are foreseen. This review is an updated report of the experimental data hitherto published in this field, as well as an introductory discussion of the most relevant problems arising in the interpretation of experimental P.E. data on d-metal compounds. Several excellent reviews have already appeared both on general aspects of P.E. spectroscopy (*1–3*), and on He(I) P.E. spectroscopy of transition metal complexes (*4*); however, research in this field is growing so rapidly that a new survey finds its justification in the presentation of updated results.

Introduction

Fundamental Process of P.E. Spectroscopy. Photoelectron (also called photoemission or photoionization) spectroscopy measures the properties of electrons ejected from a bound structure, e.g., a discrete gaseous molecule, after impact with a photon of medium or high energy, such as those produced in a inert gas discharge (~ 10–40 eV; typically 21.21 eV in He(I) $2p \rightarrow 1s$, or 40.81 eV in He(II) $2p \rightarrow 1s$), or in soft X-ray emission (up to a few thousand electron volts; most commonly used is Al K_α at 1486.6 eV). Useful observations include energy and angle distribution of the ejected photoelectrons, as well as their flux intensity or cross section of the photoionization process; by far the largest amount of available experimental data, and of interpretative work of chemical significance, deals with energy measurements, although angle and intensity data are also of potential interest and are actually considered occasionally. The photoionization processes with which we are concerned in the present review are those produced by He(I) [and occasionally also by He(II)] photons on gas-phase molecules of d-metal compounds; one can thus ionize electrons from valence orbitals rather than from inner core shells as in X-ray P.E. spectroscopy, and actually the covered region of ionization energies, up to 21.21 eV, includes all, or almost all, of the valence orbitals of most discrete molecules.

Operation in gas phase implies the obvious limitation to coordination compounds which are volatile (but this limitation is not so severe as it seems at first sight; usually a vapor pressure of $\sim 10^{-2}$ torr at the highest temperature attainable without decomposition, up to $\sim 300\,^\circ$C, is sufficient for recording a good spectrum in most existing spectrometers) and has the advantages of freedom from surface charging, surface damage, and other surface effects, often annoying in solid state P.E. spectroscopy, as well as in the good energy resolution. Since in fact gas discharge lines have very small natural widths and gaseous samples do not show phonon broadening effects, the practical limit to the resolution is set by the performance of the electron analyzer employed (usually cylindrical 127° or 180° electrostatic double-focusing detectors) and is typically in the range 30–40 meV, or 300–400 cm^{-1}, which is sufficient to detect simple vibrational structures or to resolve close-lying electronic levels.

Ionization Energies and Orbital Energies. The P.E. spectrum of a molecular species M is a display of the increase in number of ejected primary electrons versus their kinetic energy K.E. $= h\nu -$ I.E. and is a measure of the ionization energies I.E. of M, that is, of the differences between various energy levels of the M$^+$ ions, and the ground state of neutral M. For molecular structures represented by a single Slater determinant of molecular orbitals (M.O.), we can talk of ionization of single molecular orbitals, so a P.E. spectrum is strictly a measure of the ionization energies of filled molecular orbitals or of the binding energies of the electrons contained therein, which are still energy differences between states of M$^+$ and the ground state of M. Intrinsic properties of the ground state of M, such as the orbital energies, or the values of the Lagrange multipliers playing the role of pseudoeigenvalues of the M.O.s in an SCF treatment are related to the above binding energies and are numerically close to them, although not necessarily coincident with them. The first, rough approximation in this direction is that of Koopmans' theorem (5) equating the measured electron binding energies to the negative of the calculated SCF M.O. eigenvalues, provided the ground state is a closed shell:

$$\text{I.E.} = -\epsilon_{SCF}. \tag{1}$$

The sequence of observed ionization bands, or peaks, in a P.E. spectrum should therefore match the sequence of the negative of the energies of the occupied molecular orbitals, as obtained from a good SCF calculation. For instance, the ligand molecule C_6H_6 has in the region of lower I.E. shown in Fig. 1, three groups of bands at 9.25, 11.49 and 16.8 eV, assigned to the ionizations of the e_{1g} π-bonding orbital pair, of the close lying e_{2g} σ and a_{2u} π orbitals, and of the σ_{CH} bonding orbitals, respectively (6).

121

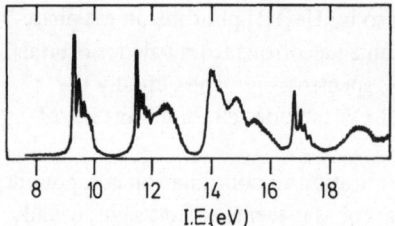

Fig. 1. He(I) P.E. spectrum of benzene (6). Like all other figures of this review, the spectrum is adapted from the reference(s) quoted in parentheses

However, Koopmans' theorem is only an approximation, which would be fully valid in case both the nuclear and the electronic configuration of the rest of the molecule remain unchanged or "frozen" during the ionization process; this is probably true of the nuclei, whose motions are slow on the time scale of electronic transitions, whereas electrons in the remaining filled orbitals undergo considerable relaxation, up to a few electron volts, so that I.E. is in practice lower than $|\epsilon_{SCF}|$ by the same amount. In several cases, including most organic molecules, relaxation effects are nearly constant in the ionization of all valence orbitals, or nearly proportional to their $|\epsilon_{SCF}|$, so an approximate form of Koopmans' relation

$$- \epsilon_{SCF} = a + b \ (\text{I.E.}) \tag{2}$$

(with $b \sim 1$) still holds, thus allowing qualitative correlations for assignment purposes, as well as for semiquantitative comparisons. Unfortunately this is not the case with d-metal compounds, where relaxation following d-orbital ionization is usually much larger than that following ionization of ligand-based orbitals (7–12), leading to situations where the ionization energies of d-orbitals may be smaller than those of ligand orbitals, yet their orbital energies are not the highest ones. Koopmans' theorem is therefore of much more limited value with transition metal compounds both because open-shell structures are more common in their ground state, and because of different relaxation on different orbitals, so that assignment has to be aided by other means, such as induction and comparison between compounds of related structure, or variations in P.E. band intensities between He(I) and He(II) irradiation. Cross sections for photoionization depend in fact both on the nature of the ionized orbital and on the energy of the irradiating photons, the transition matrix element being related to the dipole- or velocity-weighted overlap between the ionized bound orbitals and the free wave of the ejected electron. Although simple generalized calculations are not available, it is experimentally known that d-orbital ionizations exhibit a marked relative increase in intensity on passing from He(I) to He(II) excitation, a fact which is often used for diagnostic purposes (13–18).

d-Ionizations in P.E. Spectra. Bearing in mind these difficulties, it is nevertheless possible to arrive at identification of d-orbital ionizations in the P.E. spectra of transition-metal compounds. Thus, VCl_4 has a spectrum similar to $TiCl_4$, except for an extra band at 9.41 eV, obviously related to the only difference in electronic structure between the two compounds, i.e., the presence of an extra electron in a d-like orbital in VCl$_4$ (*19*) (Fig. 2). Under favorable circumstances, therefore, P.E. spectroscopy allows identification of d-levels and yields informations which either reinforce or extend those afforded by absorption spectroscopy data as interpreted by means of ligand-field theory. Obvious differences are that empty d-orbitals escape direct observation in P.E. spectroscopy; thus the d^3 configuration of Cr(acac)$_3$ gives only one P.E. signal [ionization to $d^2(^3T_{1g})$] instead of two or three $t_{2g} \rightarrow e_g$ transitions observed in electronic absorption spectroscopy for octahedral Cr(**III**) complexes. Conversely, filled systems like d^{10} in the $e^4 t_2^6$ configuration of tetrahedral Ni(CO)$_4$, which do not show intrasubshell transitions in electronic absorption, show in P.E. spectroscopy well-defined ionization patterns with a 2:3 intensity ratio in the $\rightarrow e^3$ and $\rightarrow t_2^5$ bands (*11, 20*) (Fig. 3). Again, it must be borne in mind that the customary interpretation of corresponding quantities is different in electronic absorption spectroscopy (hence also in conventional ligand-field theory) and in P.E. spectroscopy: In the case of Ni(CO)$_4$, the energy difference $9.76 - 8.93 = 0.83$ eV between the binding energies of dt_2 and de electrons plays the role of a Δ value but is not exactly the same Δ value which we would define in a conventional ligand-field structural assignment based on absorption-spectroscopic considerations (apart from the fact that a $de \rightarrow dt_2$ transition is not directly observable in a d^{10} system), both because of possible different relaxation contributions to the $\rightarrow e^3$ and $\rightarrow t_2^5$ ionizations and because the observed P.E. energy difference of 0.83 eV is between two M$^+$ states and is therefore a Δ value

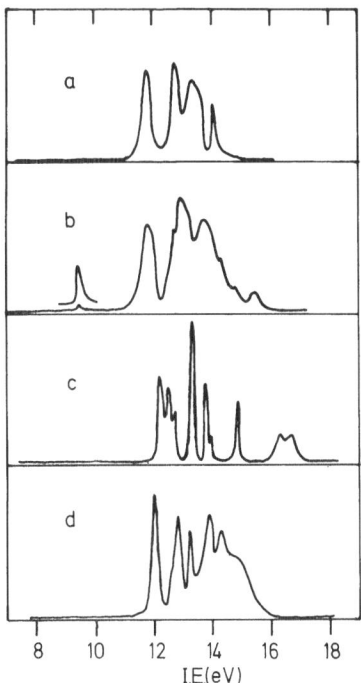

Fig. 2. He(I) P.E. spectra of tetrahedral halo and oxospecies: (a) TiCl$_4$ (*19*); (b) VCl$_4$ (*19*); (c) RuO$_4$ (*70*); (d) CrO$_2$Cl$_2$ (*70*)

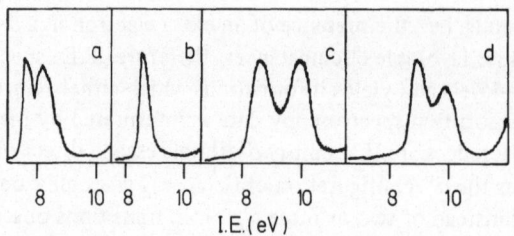

Fig. 3. Low-energy part of the He(I) P.E. spectra of metal carbonyls: (a) $V(CO)_6$ (*21*); (b) $Cr(CO)_6$ (*28*); (c) $Fe(CO)_5$ (*20*); (d) $Ni(CO)_4$ (*20*)

of Ni(I) in $[Ni(CO)_4]^+$ (and in a particular nonequilibrium nuclear configuration of the molecular ion, i.e., with the nuclei in an unrelaxed position, equal to that of equilibrium in the ground state neutral molecule) rather than Δ of Ni(O) in $Ni(CO)_4$. The two quantities are, however, grossly related and would coincide in the extreme Koopmans limit.

P.E. Spectroscopy, Electronic Absorption Spectroscopy, and Ligand Field Theory. Within the above-discussed limitations, P.E. spectroscopy affords a picture of ligand field effects which is related to, even if not identical with, that afforded by electronic absorption spectroscopy and lends the same kind of support or evidence for the expectations of ligand field theory; thus, we observe symmetry-determined ligand-field splitting patterns of d-levels (Fig. 4) [see below, e.g., octahedral $Cr(CO)_6$ (*20–22*), trigonal-bipyramidal $Fe(CO)_5$ (*20, 23*), or tetrahedral $Ni(CO)_4$ (*11, 20*)], although energy resolution is generally poorer than in absorption spectroscopy and lower symmetry and other finer effects are more difficult to appreciate in P.E. spectroscopy. Russell-Saunders coupling effects are evident particularly in the ionization of

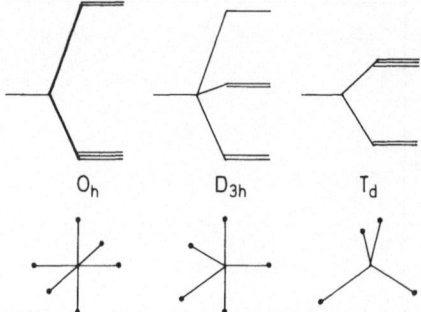

Fig. 4. Schematic ligand-field splitting patterns of d-orbitals in various coordination symmetries

open shell neutral molecules, whose molecular ions exhibit coupling between two open shells produced on ionization, or multiplet structure in the ionized open shell such as in $V(CO)_6$ $(t_{2g})^5$ which becomes $(t_{2g})^4$ $(= {}^3T_{1g} + {}^1T_{2g} + {}^1E_g + {}^1A_{2g})$ in $[V(CO)_6]^+$ (21); as a general rule, to a first approximation the P.E. spectrum shows energy splitting and intensity ratio between components, which correspond to the splitting patterns and overall multiplicity of the Russell-Saunders coupling terms in the molecular ion (21). Spin-orbit coupling effects are also evident in some cases, e.g., in the comparison between homologous compounds containing light, respectively heavy atoms, e.g., in $Mn(CO)_5 I$ as compared with $Mn(CO)_5 Cl$ (15) or in substituted rhenium versus manganese carbonyls, $Re(CO)_5 X$ and $Mn(CO)_5 X$ (15). Vibrational structure in the molecular-ion states can also in principle be observed and helps in assigning the bonding, nonbonding, or antibonding character of the ionized molecular orbital; however, in the majority of cases it appears unresolved in molecules of relatively high complexity such as those of transition metal compounds. As a consequence, vertical rather than adiabatic ionization energies are usually reported for metal compounds.

P.E. spectroscopic data on d-shells differ markedly from absorption spectroscopic data in giving absolute values of energy levels (referred to the vacuum level) rather than relative distances among the split components of the manifold of d-orbitals, and in yielding immediate comparative evidence for the existence and energy position of all other valence orbitals, including delocalized coordinative orbitals and ligand orbitals besides the mainly metal-localized ones. This yields additional, and more complete, information on the electronic structure of the molecule as a whole, not restricted just to the d-shell. It becomes evident then that d-orbital energies become increasingly negative with increasing atomic number Z along a transition period, so that d^n ionizations occur, for $n = 1, 2, 3$ at low I.E. ($\sim 6-9$ eV) and are usually well separated from ionizations of ligand-based orbitals in compounds of early transition metals, e.g., Ti to Cr, while d-ionizations in late transition metal complexes, e.g., d^9 in Cu(II) complexes or d^{10} in Zn(II), Ag(I) or Hg(II) complexes, occur at higher I.E. ($\sim 10-18$ eV) and are intermixed or masked by neighboring bands of ligand orbitals; intermediate cases and alternative behavior can also occur, e.g., in d^7 of Co(II) and d^8 of Ni(II) compounds, thus adding to the ambiguity of assignment, as will be discussed in the subsequent sections.

When d-ionizations can be sharply differentiated from ligand ionizations, it is evident that the energy differences between the two types of processes are usually smaller than the differences of transition energies from optical ligand-field and charge-transfer spectra, leading often to the conclusion that d-orbital energies are lower than those of coordination bonds or of ligand orbitals, even if the former are more easily subject to optical excitation in electronic absorption spectroscopy. This situation is not unexpected in absorption spectroscopy, where it is explained by the substantial contribution of interelectronic repulsion energies to the transition energies, besides the one-electron energy terms; however, absorption spectra do not afford commonly direct evidence for such inequalities, whereas P.E. spectra afford currently immediate evidence.

Along the same line, other structural effects, such as trends in atomic charge distributions, inductive effects, and π-donation effects can be seen directly in P.E. spectra of d-metal compounds, while electronic absorption spectroscopy can yield at most indirect or ambiguous evidence for the same facts. Just to mention one dramatic example, dialkylamido ligands in $V(NR_2)_4$ render vanadium so negative by N $\rightarrow$ V π-donation that d-orbitals are easily ionized at 6.2 eV before all other orbitals (lone pairs of N at 7.08, 7,60, and 8.28 eV) (24), whereas the analogous disilylamido ligands in $Cr[N(SiR_3)_2]_3$ transfer negative charge to silicon via N $\rightarrow$ Si(d) π-donation, thus leaving chromium more positive and its d-orbitals low enough in energy so as not to give distinct ionization peaks before those due to the lone pairs of N at 8.07 and 8.76 eV (25). The electronic absorption spectrum of the latter compound is instead normal and comparable to those of other Cr(**III**) complexes: in fact, d-orbitals are the first to undergo spectroscopic excitation (of ligand-field type) (26); therefore, electronic absorption spectroscopy does not reveal the large difference existing between the electronic structure of alkylamido and silylamido complexes.

Having thus anticipated in this section the main general aspects of P.E. spectroscopy of volatile d-metal compounds, including assignment problems and the cautions to be taken in interpretation, we shall now proceed to a survey of the experimental data available from the literature up to the beginning of 1977. We divide the treatment into two sections, the first comprising complexes of metals in zero or low oxidation state (carbonyls, arenes, cyclopentadienides, etc.) and the second containing data on complexes of metals in higher, "normal" oxidation states with more electronegative ligands. Such classification is intended for practical purposes only and is to a good extent arbitrary, except for the convenience of carrying the analysis of P.E. spectra by classes of ligands and for the clearer definition of d-orbital ionizations at low I.E. values in the former type of complexes, as will be discussed in detail.

Compounds of d-Metals in Low Oxidation States

Neutral ligands which stabilize low oxidation states of transition metals, such as CO, PF_3, olefines and arenes, usually exhibit as free molecules relatively high ionization energies ($\gtrsim$ 11 eV), which are further increased as the corresponding orbitals become involved in coordinative dative bonds. Aromatic ligands have a low ionization energy (e_{2g} in benzene) typically around 9.2 eV, which is however easily identified in the corresponding metal complexes, hence clearly separated from neighboring ionizations in the same energy region. As a result, the region of low ionization energies ($\lesssim$ 11 eV) allows separate (or nearly so) observation of ionization of the mainly metal-localized d-type orbitals. Taking as a rough reference the first atomic ionization of the metals ($\sim$ 6–9 eV for 3 d elements) and even allowing for the facts that nd is ionized in the complexes before $(n + 1)s$, that atomic orbitals are strongly deformed in complexes, and that the true atomic charge may vary around zero, ionization of metal-centered orbitals can be expected and is actually observed, in the region $6-11$ eV, where P.E. ionization patterns should reproduce the ligand field splitting patterns of the orbital energy levels of the partly filled shell of predominant d-character, apart from the already mentioned limitations due to possible deviations from Koopmans' behavior. Such easy observation of the partly filled d-shell of low-valent transition metal compounds by UPS contrasts to and is in some way complementary to the situation occurring in visible-UV electronic absorption spectroscopy, where low-energetic intense charge transfer transitions often mask $d-d$ transitions and may make direct observation of the d-level system virtually impossible. UPS affords therefore a neater and more complete way of observing experimentally the ligand-field splitting patterns of d-orbitals and of checking the fundamental aspects of the corresponding theoretical models. In addition, a number of finer structural effects can be observed and interpreted, such as multiplet splitting in open shell molecules, spin-orbit coupling in M.O.s of the partly filled shell, lower-symmetry effects, mixing of d-orbitals with possible ligand orbitals of the same symmetry, inductive effects on the molecular charge distributions, and intensity or hyperchromic effects.

Metal Carbonyls (Unsubstituted). (Table 1). The UPS He(I) spectrum of free CO shows three well-defined bands assigned, by general agreement, to ionization of the practically nonbonding orbital $\sigma_g\, 2p$ (narrow peak with vertical ionization potential (I.P.) of 14.01 V), π-bonding $\pi_u\, 2p$ (series of peaks with maximum at 16.91 eV), and $\sigma_u\, 2p$ at 19.72 eV (27). The same, or strictly related ionizations occur in metal carbonyls in approximately the same energy regions; actually the "lone pair" σ_g orbital would be expected to move to lower orbital energies as it becomes σ-bonding in the complexes, but the effect may be contrasted or even reversed by π-backdonation effects which make the whole CO moiety more negative. Furthermore, crowding of several CO groups on the same metal center leads to interligand interactions, hence to doubling and shifting of levels, and results eventually in a broad, unshaped photoionization band extending over the whole region between ~ 13 and 19 eV, often with maxima between 15.5 and 16.0 eV, and shoulders or side bands at ~ 13.5 and 17.5 or 18.0 eV. Thus, for example, $Cr(CO)_6$ has a band at 13.38 eV, followed by a broad envelope between 14 and 16.5 eV ($20-22, 28$). Little or no structure can be identified under such broad band envelopes, which account as a whole for all M.O.s deriving from σ_g, π_u and possibly also σ_u of the CO ligands; low resolution prevents more detailed assignments, and ionizations in this region are in practice not very informative as to the structure of metal carbonyls.

Therefore, ionization events occurring at lower I.P. appear more interesting since they involve the d-type metal-centered orbitals and correspond to the d^n configurations in the ligand-field splitting patterns appropriate to the molecular symmetry of $M(CO)_n$, Thus we have one ionization band with multiplet structure, at 7.52 and 7.88 eV, from the t_{2g}^5 configuration of octahedral $V(CO)_6$ (21), one narrow band at 8.42 eV from t_{2g}^6 in octahedral $Cr(CO)_6$ (22), two bands of equal intensity at 8.60

127

Table 1

Compounds	Formal metal electronic configuration	Ionization energies (eV)[a] — d	Other	CO	References[b]
V(CO)6	d^5	17.52		12.95 13.64 17.3	21
Cr(CO)6	d^6	8.40	7.88[c]	13.38 14.21 14.40 15.12 15.60 16.2 17.82	20, 21, 22, 28
Cr(CO)5NH3	d^6	7.56[f]	7.85[g]	[12.0–16.0][h] 17.6	35
Cr(CO)5NMe3	d^6	7.45[f]	7.76[g]	[12.1–15.5][h] 16.7	35
Cr(CO)5PH3	d^6	(7.90)[f]	8.03[g] 10.57[i]	]12.1–15.5][h] 17.6	35
Cr(CO)5PMe3	d^6	(7.58)[f]	7.72[g] 11.43[i]	[12.5–16.0][h] 17.3	35
Cr(CO)5CNMe	d^6	(7.61)[f]	7.77[g] 10.00[i]	[11.2–15.5][h] 16.9	35
Mo(CO)6	d^6	8.50		13.32 14.18 14.41 14.66 15.2 15.6 17.71	20, 22, 28, 29
W(CO)6	d^6	8.56		13.27 14.20 14.42 14.88 15.2 15.54 17.84	20, 22, 28
W(CO)5NH3	d^6	[17.54 7.75[f]]	8.06[g]	[12.0–16.0][h] 17.0	35
W(CO)5NHMe2	d^6	[17.41 7.62[f]]	7.95[g]	[12.3–15.8][h] 17.0	35
W(CO)5NMe3	d^6	[17.41 7.62[f]]	7.96[g]	[12.3–15.8][h] 16.8	35
Mn(CO)5H	d^6	8.85[f]	9.14[g] 10.55[i]	[13.9–17.0][h]	12, 36, 37, 38, 39
MnCl(CO)5	d^6	10.56[f]	11.18[g] 8.94[j] 9.56[k]	14.0 [14.4–15.2][h] 17.0	37, 39, 40, 15
MnBr(CO)5	d^6	10.11[f]	10.81[g] 8.86[j] 9.56[k]	13.8 [14.3–15.4][h] 17.4	39, 40, 15
Mn(CO)5I	d^6	9.69[f]	10.44[g] 8.44 8.74[j] 9[k]	13.7 [14.1–15.0][h] 17.6	39, 40, 15
Mn2(CO)10	d^6	8.02[f]	[8.35 9.03[f]] 9.49[i]	[13.0–16.6][h] 17.7	15, 39
Mn(CH3)(CO)5	d^6	8.65[f]	9.12[g] 9.12[j]	12.0 [12.6–13.8][h] 16.0	12, 37, 38, 39, 44, 45
Mn(CF3)(CO)5	d^6	9.20[f]	10.30[g] 10.75[i]	13.5 [14.0–15.4][h] 17.6	28, 39, 44, 45
Mn(CO)5(SiH3)	d^6	8.99[f]	9.38[g] 11.9[m]	[13.7–17][h]	41
Mn(CO)5(SiF3)	d^6	~9.8	10.4[i]	[13.5–17][h]	42
Mn(CO)5(SiMe3)	d^6	9.0[f]	9.3[g] 10.8[i]	[13.5–17][h] 13.1[n]	41
Mn(CO)5(SiCl3)	d^6	(19.36) 8.90[f]	9.58[o] 11.36[p] 12.17 12.92	[15.1–17][h] 14.32[j] 13.71	15
Mn(CO)5(GeH3)	d^6	9.26[g]	11.54 11.59	[13.4–17][h]	41
Mn(CO)5(COCF3)	d^6	8.63[f]	(9.01)[g] 10.21[g] 9.66[f]	12.2 [13.3–15.0][h] 17.6 14.30	39
Mn(CO)5(SnMe3)	d^6	9.61[f]	8.26[j]	(15.03) (16.02)[s] 13.63	15
MnBr(CNMe)(CO)4	d^6	9.15[f]	8.86[j]	(15.03) [13.5–17] 8.86	40
Re(CO)5H	d^6	19.46	9.53[g]	[13.5–17][h] 10.5[i]	15, 36, 41
ReCl(CO)5	d^6	19.09	9.91 10.83	[13.7–17.5][h] 11.23[t]	15, 36, 110
ReBr(CO)5	d^6	18.83	9.94 10.52	[13.6–17.5][h] 10.90[t]	15, 36, 110
Re(CO)5I	d^6	18.36	9.71 10.02	[13.6–17.8][h] 10.44[t]	15, 36, 110
Re2(CO)10	d^6	8.07[l] 18.57 8.86 9.27	9.58[f]	[13.0–17.3][h] 10.5[i]	15, 36
Re(CH3)(CO)5	d^6	18.71	8.93 9.51[o]	~15.0 12.8[n]	15, 36
Re(CO)5(SiH3)	d^6	18.9 9.11[f] 9.5	9.6[g]	[13.6–17.0][h] 11.6[i]	41
Re(CO)5(GeH3)	d^6	18.9 9.13[f] 9.4	9.6[g]	[13.6–17][h] 11.4[i]	41
Re(CO)5(COCF3)	d^6	19.40 9.69[f]	9.97[g]	10.80	36
Fe(CO)4H2	d^8	9.65	9.86[v]	[14–16][h] 10.95 11.30[i]	12
Fe(CO)5	d^8	8.60[u] 8.61[g]	9.6[w]	>13.5 10.5 12.3[y]	23, 20
Fe(CO)4(C2H4)	d^8	8.4 9.2[l]	9.8[x]	>13.5 10.5 12.3[y]	23

Top section

Compound	d^n	Ionization potentials (eV)	References
Fe(CO)$_3$(ClCH$_2$]$_3$)	d^8	9.26 [18.62] 11.07 12.11 12.57 13.04][z] 13.88 14.28 15.27 17.60 20.78	18, 46
Fe(CO)$_3$(C$_4$H$_6$)	d^8	8.82 [(8.23)] 9.93 11.52 12.94][z] [14–16][h]	14, 47, 118
Fe(CO)$_3$(C$_4$H$_5$Me)	d^8	7.84[a']	47
Fe(CO)$_3$(C$_4$H$_4$)	d^8	(8.17) 8.45 9.21[y] (12.81) 13.69 14.2 16.99 17.46 20.31	17, 47
Fe(CO)$_3$(C$_4$H$_3$COMe)	d^8	8.27[a']	47
Fe(CO)$_3$(C$_4$H$_3$CHO)	d^8	8.32[a']	47
Fe(CO)$_3$(C$_4$H$_3$NH$_2$)	d^8	7.77[a']	47
Fe(CO)$_3$C$_6$H$_8$	d^8	(7.98) 8.56 9.33[g] 11.04[h] 12.17[i]	118
Fe(CO)$_3$(C$_7$H$_{10}$)	d^8	(7.78) 8.46 9.12[g] 10.86[h] 11.71[i]	118
Fe(CO)$_3$(C$_8$H$_{12}$)	d^8	(7.45) 8.27 8.87[g] 10.44[h] 10.87[i]	118
Fe(CO)$_3$C$_7$H$_8$	d^8	(7.76) 8.39 10.23[e] 11.10[h] 11.82[i]	118
Fe(CO)$_3$(C$_8$H$_8$)	d^8	(8.78)[g] 8.74[d'] 10.6[f] 11.63[h]	118
Ru(CO)$_3$(C$_6$H$_8$)	d^8	7.84[d'] 8.01 (9.39)[g] 11.01[h] 11.83[i]	118
Ru(CO)$_3$(C$_7$H$_{10}$)	d^8	7.96 (9.40)[g] 10.84[h] 11.64[i]	118
Co(CO)$_4$H	d^8	8.90[v] 9.80[v] 11.5[i] [13.8–17][h]	41
Co(CO)$_4$(SiH$_3$)	d^8	8.85[u] 9.90[v] 11.9[m] [13.8–17][h]	41
Co(CO)$_4$(GeH$_3$)	d^8	8.80[v] 9.90[v] 11.9[q]	41
Ni(CO)$_4$	d^{10}	8.90[b'] 9.77[f] 14.12 14.95 15.7 18.25	11, 20
Co(CO)$_3$(NO)	d^{10}	8.90[b'] 9.82[f] 14.53 14.92 18.17	11, 20
Fe(CO)$_2$(NO)$_2$	d^{10}	[8.56] 8.97[b'] 9.74[f] 14.71 15.43 16.27 18.20	11, 20

Bottom section

Compound	d^n		σ M–H (a₁)	σ M–P	σ M–H	σ M–P	Fluorine lone pair	References
PF$_3$							15.91 17.45 18.54 19.42	29, 30, 31, 32
Cr(PF$_3$)$_6$	d^6	9.29				12.31[c]	15.80 17.36 19.3	33, 111
Mo(PF$_3$)$_6$	d^6	9.17				12.84	15.80 17.36 19.1	29, 33
W(PF$_3$)$_6$	d^6	9.30		12.94	13.93	13.48	15.85 17.44 18.7	33
MnH(PF$_3$)$_5$	d^6	9.47	11.30	12.26	13.52	12.64 12.93	15.85 17.43 19.4	33
Fe(PF$_3$)$_5$	d^8	9.15[u] 10.43[v]				13.08	15.83 17.24 19.1	33, 111
Ru(PF$_3$)$_5$	d^8	9.17[u] 11.07[v]		12.8		13.25	15.75 17.18 18.9	33
CoH(PF$_3$)$_4$	d^8	9.58[f] 10.56[f]	12.12		13.65	13.25	16.00 17.46 19.4	33, 111
RhH(PF$_3$)$_4$	d^8	9.70[f] 11.79[f]				13.83	15.90 17.42 19.3	33
IrH(PF$_3$)$_4$	d^8	9.82[f] 11.95[f]				14.18	16.01 17.42 19.4	33
Ni(PF$_3$)$_4$	d^{10}	9.55[b'] 10.58[f]				13.09	15.83 17.43 19.27	29, 30, 31, 32, 33, 34
Pd(PF$_3$)$_4$	d^{10}	9.9[b'] 12.2[f]				13.7	15.84 17.4	33, 34
Pt(PF$_3$)$_4$	d^{10}	9.8[b'] 12.3[f]				14.5	15.8 17.4 19.5	30, 32, 33, 34

a Shoulders in parentheses.
b Numerical data are taken from the references in italics
c $^3T_{1g} + {}^1T_{2g} + {}^1E_g + {}^1A_{1g}$.
e e-type sublevel.
g b$_2$-type sublevel.
h Broad envelope.
i M–L bonding.
j e(p$_\pi$ halogen).
l a$_1$(p$_\sigma$ halogen).
m σ Si–H.

n σ C–H.
o e(''d'$_2$) + b$_2$(''d'$_2$) + σ M–L.
p σ Si–Cl.
q σ Ge–H.
r e(σ Sn–C).
s a(σ Sn–C) + σ C–H + CO.
t e(''d'$_2$) + b$_2$(''d'$_2$) + e(p$_\pi$ halogen) + a$_1$(p$_\sigma$ halogen).
u e'-type sublevel.
v a$_2$-type sublevel.
w a$_2$-type sublevel.
x b$_1$-type sublevel.

y L orbitals.
z ''d'' + L orbitals.
a' adiabatic.
b' t$_2$-type sublevel.
c' P lone pair
d' mixed with π(diene)
e' π of uncomplexed C=C
f' π(diene)
g' π(1a$_2$ of the C=C–C=C moiety)
h' π(1b$_1$ of the C=C–C=C moiety)
i' σ(diene) + CO

and 9.86 eV in trigonal bipyramidal $Fe(CO)_5$ having a $d^8 (e'')^4 (e')^4$ configuration, and two bands at 8.90 and 9.77 eV, with intensity ratio $3:2$ in $Ni(CO)_4$, corresponding to ionization of the completely filled tetrahedral sublevels $(e)^4 (t_2)^6$ (20). $V(CO)_6$ is the only open-shell system in this series, and multiplet splitting occurs in the t_{2g}^4 configuration of the molecular ion according to the usual Russell-Saunders coupling scheme; however, the expected splitting is not completely resolved, and of the predicted terms $^3T_{1g}$, ^1E_g, $^1T_{2g}$, and $^1A_{1g}$, the last one is not observed separately, while the other two singlets are predicted at nearly equal energies. The proposed assignment is therefore $^3T_{1g}$ as the maximum at 7.52, and $(^1T_{2g} + {}^1E_g)$ as the shoulder at 7.88 eV. The observed intensities are close to the expected $9:5$ ratio (21). $Mo(CO)_6$ and $W(CO)_6$ (20, 22, 28, 29) are very similar to $Cr(CO)_6$ in their UPS spectra, with small differences in the I.E. of the t_{2g}^6 electron system; spin-orbit splittings are small and not resolved, even in the tungsten compound.

The d-ionization patterns are more complex in binuclear carbonyls. For the $14d$ electrons of $Mn_2(CO)_{10}$ and $Re_2(CO)_{10}$, P.E. spectra clearly indicate (15) an orbital energy sequence of three e levels (the two lowest ones being very close in energy) probably containing the $\pi(xz, yz)$ and $\delta(xy)$ orbitals, followed by a nondegenerate orbital, possibly the M–M bonding $\sigma(d)$, all of these levels being contained in a relatively narrow interval of 1.0–1.5 eV.

Trifluorophosphine Metal Complexes can be compared with metal carbonyls, and the similarity between PF_3 and CO as ligands had been noted before. Free PF_3 has in its P.E. spectrum a band at 12.31 eV (P lone pair), followed by bands between ~ 15 and 20 eV due to the π_F lone pairs grouped into four levels $a_1 + a_2 + 2e$ and, at still lower orbital energies, to the three P–F σ bonds $(e + a_1)$ (29–32). In the neutral metal complexes of PF_3, the P lone pair becomes σ-bonding to the metal, hence shifted to higher I.E. (between ~ 12.7 and 14.5 eV), while the ionization patterns in the region of lower i.e. strongly resemble those of the metal carbonyls of the same d^n configuration and geometric structure, apart from a shift of I.E. to higher values because of the better electron-acceptor properties of PF_3 if compared to CO. Thus, we have one peak at 9.3 eV from the t_{2g}^6 configuration of octahedral $Cr(PF_3)_6$ (33), two peaks of equal intensity at 9.15 and 10.43 eV from $(e'')^4 (e')^4$ of $Fe(PF_3)_5$ (33), and two peaks $(3:2)$ from $(e)^4 (t_2)^6$ of tetrahedral $M(PF_3)_4$ (M = Ni, Pd, Pt) (29–34). In the latter series, $\Delta_{tetr.} = E(t_2) - E(e)$ increases in the order $Ni(1.05) <$ $Pd(2.3) < Pt(2.5)$ eV, possibly as a consequence of increasing extent of σ donation effects.

In the mixed hydride-trifluorophosphine complexes $HM(PF_3)_4$ (M = Co, Rh, Ir) and $HMn(PF_3)_5$, the P.E. spectral patterns reflect an analogous bond situation, with d-ionizations between ~ 9 and 11 eV, and M–H ionizations around 11–12 eV [e.g., 11.30 eV in $HMn(PF_3)_5$ (33)]. σ_{M-P} ionizations occur again around 13–14 eV, with no evident splitting (33). These data have been interpreted as an indication of a strong σ donor ability of the PF_3 ligand (33), possibly larger than that of CO. On the whole, PF_3 is obviously more electron-attracting than CO [compare, for example, the d^6 ionization at 8.40 eV in $Cr(CO)_6$ (22), shifted to 9.29 in $Cr(PF_3)_6$ (33)], although the role of π bonding is still subject to controversy, and conflicting conclusions are suggested from other investigation techniques (33).

Substituted Metal Carbonyls. While repeating the basic features of the UPS spectra of the parent unsubstituted carbonyls of the same d^n configuration and c.n., the spectra exhibit here a number of additional features pertaining to lower-symmetry splitting of the d-levels, to the ionizations of orbitals mainly localized on the substituent ligands or on their bonds to the metal, and to the inductive effects exerted by such ligands. The latter mentioned ionization processes occur often at energy values close to those of d ionizations, and the involved orbitals may be of considerably mixed composition, a situation which may lead to ambiguities and has indeed given rise to controversies in the assignments. We shall briefly review the most significant examples, classified according to the d^n configuration to which the metal-centered partly filled shell can be approximated.

Monosubstituted octahedral carbonyls of d^6 type include two widely investigated groups, $M(CO)_5L$ (M = Cr, W; L = neutral ligand) and $M'(CO)_5X$ (M' = Mn, Re; X = negative ligand). In $Cr(CO)_5(NH_3)$, $Cr(CO)_5(PH_3)$, and analogous complexes (35), the P.E. spectrum differs from that of $Cr(CO)_6$ in three main respects:

i) The d^6 ionization is a well split doublet with intensity ratio 2:1 [e.g., 7.45 and 7.76 eV in $Cr(CO)_5(NMe_3)$] according to the low-symmetry splitting $t_{2g}(O_h) \rightarrow b_2 + e(C_{4v})$, with $b_2 < e$;

ii) the center of gravity of d-ionizations is shifted to lower I.P. [compare 8.40 eV in $Cr(CO)_6$] as a consequence of higher negative charge on the metal atom because of replacement of one CO molecule by a less effective π-acceptor ligand;

iii) appearance of an extra band due to $\sigma(M-L)$ ionization, e.g., $\sigma(Cr-P)$ at 11.43 eV in $Cr(CO)_5(PH_3)$ (35). Tungsten derivatives behave in the same manner and also exhibit some not well-resolved spin-orbit splitting in the $e(d)$ ionization band.

$HMn(CO)_5$ is the simplest structure among d^6 substituted manganese pentacarbonyls and exhibits a double band at 8.85 and 9.14 (2:1) and a less intense band at 10.55 eV, accounting altogether for the d^6 $(e + b_2)$ and for the $\sigma_{Mn-H}(a_1)$ ionizations. There is now general agreement that the double band is due to $(e + b_2)$ with a small low-symmetry splitting, while σ_{Mn-H} is assigned as the band at 10.55 eV (12, 36), but the problem was subject to considerable controversy in earlier reports (37–39), the uncertainties arising from poor experimental resolution of the band components at 8.85 and 9.14 eV, from the cross section of ionization of hydride orbitals being larger and not comparable to that of metal orbitals, and from a possible inversion with respect to Koopmans' behavior. $HRe(CO)_5$ is assigned in a similar way, with a spin-orbit split doublet at 8.86–9.15 eV as $e(Re)$ and peaks at 9.53 as $b_2(Re)$ and 10.5 as $\sigma(Re-H)$ (36, 41). Also $H_2Fe(CO)_4$ can be considered here; it has an unresolved $(e + b_2)$ band at 9.65 eV, while the two σ_{Fe-H} bonds give rise to two distinct ionizations, at 10.95 and 11.30 eV (12).

In $Mn(CO)_5X$ (X = Cl, Br, I, CH_3, and other group IV YH_3 and related substituents) up to six orbitals, grouped into four levels, are ionized in the region of low I.P., including $d^6(e + b_2)$, $\sigma_{Mn-X}(a_1)$ and the π-type lone pairs of $X(e)$, or the e combination of H–Y σ bonds (Fig. 5). Thus, for example, $Mn(CO)_5Cl$ has a sequence of bands (or band components) at 8.94, 9.56, 10.56, and 11.18 eV in approximate intensity ratio $2:(\lesssim 1):2:1$. Again, the actual assignment is controversial: (40) and (37) suggest the 8.94 and 9.56 eV bands to be due to $n_\pi(Cl)$ (e) + $\sigma_{Mn-Cl}(a_1)$ and the 10.56 and 11.18 eV bands to be the split components $e + b_2$ of the d^6 manganese configuration; Higginson and co-workers (15) prefer to regard the band at 9.56 as b_2 and 11.18 as a_1, also on the ground that He(II) spectra show increased relative intensity of the two bands of lowest I.P., which should therefore correspond to orbitals of predominant metal-d character. The situation is further complicated by the difficult observation of the intermediate nondegenerate orbital, whose ionization band is often of small intensity, as in $Mn(CO)_5Cl$, or not evident at all, as in $Mn(CO)_5I$ (15, 39, 40) and $Mn(CO)_5(SiH_3)$ (41). However, in $Mn(CO)_5(SiF_3)$ the inductive effect of fluo-

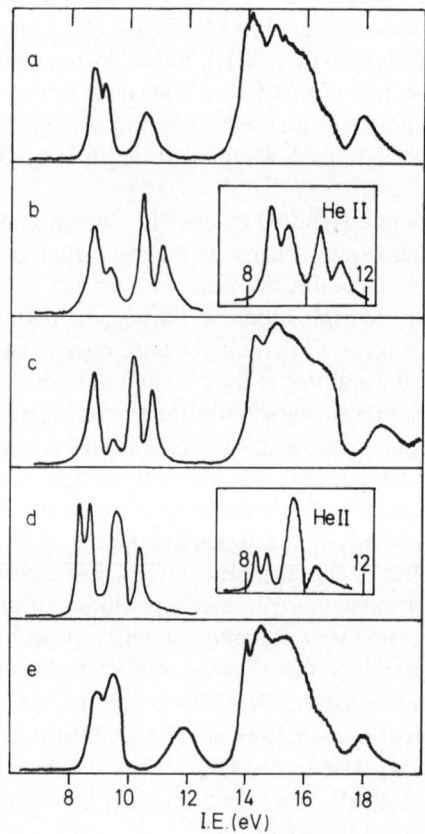

Fig. 5. He(I) [and in part He(II)] P.E. spectra of substituted manganese carbonyls: (a) $Mn(CO)_5H$ (41); (b) $Mn(CO)_5Cl$ (15); (c) $Mn(CO)_5Br$ (15); (d) $Mn(CO)_5I$ (15); (e) $Mn(CO)_5(SiH_3)$ (41)

rine shifts the a_1 band at 10.4 eV out of the high I.P. side of the unresolved $(e + b_2)$ band peaked around 9.8 eV *(42)*. In iodine derivatives and in the corresponding substituted rhenium derivatives, there are additional spin-orbit splitting effects whose theory has been reviewed *(36–43)*. It is, however, to be recalled that no simple assignment to bond orbitals can be made in cases where the same irreducible representation occurs more than once among the M.O.s in a narrow energy interval, and *Hall (36)* suggest convincingly that the two e ionizations in $Re(CO)_5X$ arise from levels of mixed $\pi(X)-d(Re)$ composition, halogen character being predominant in the e level of higher I.P. for Cl, in the e level of lower I.P. for I, and shared to a comparable extent between both e levels for Br. The most widely accepted ordering of I.P.s for the highest occupied M.O.s of $M(CO)_5X$ species (M = Mn, Re; X = Cl, Br, I) can therefore be formulated as $e(\pi_X, d) < b_2(d) < e(\pi_X, d) < a_1(\sigma_{MX})$, apart from spin-orbit effects; however, the problem cannot yet be considered to be definitely settled.

For X = CH_3, SiH_3, GeH_3, and similar ligands, the low-energy ionization patterns of $M(CO)_5X$ (M = Mn, Re) are to some extent similar to those of the corresponding mixed halide-carbonyls, and metal-d e and b_2 bands have been assigned in all cases (except when they coalesce in a single broad band); σ_{MX} ionizations are often not evident, while σ_{CH}, σ_{SiH}, and other analogous ionizations are more easily identified, usually at higher I.P., just before onset of the broad CO ionization bands. Controversial assignments are, however, reported in some cases. As an example, we shall consider $M(CO)_5(CH_3)$ (M = Mn, Re). The manganese compound has only one broad band, around 9 eV, originally thought to contain only the $(e + b_2)$ metal orbitals *(39)* under two components, at 8.46 and 9.10 eV in intensity ratio 1:2, i.e., in a inverted order for which a possible interaction with π^* acceptor orbitals of CH_3 was invoked *(39)*. As a matter of fact, the broad ionization band can be resolved under better experimental conditions into three components at 8.65, 9.12, and 9.49 eV [see *(12, 44, 45)*] in the same sequence of I.P. as $e(Mn) < b_2(Mn) \lesssim a_1(\sigma_{Mn-C})$, overcoming a previous assignment $a_1 < e < b_2$ *(37, 38)*. The spectrum of $Re(CO)_5(CH_3)$ is somewhat similar, showing a first band, actually due to two components at 8.72 and 8.98 eV, whose splitting of 0.26 eV, close to the expected ξ_{Re} value, suggests assignment to $e(Re)$, and a second band at 9.53 assigned as the unresolved superposition of $b_2(Re) + a_1(\sigma_{Re-C})$ *(36)*. Series of analogous $M(CO)_5(YH_3)$ complexes (M = Mn, Re; Y = C, Si, Ge) have been also investigated *(41, 42)*, where small energy shifts in metal-d ionizations have been related to a scale of σ-accepting power $SiH_3 >$ $GeH_3 > CH_3$, whereas π donor effects M $\rightarrow$ Y should be negligible *(41)*.

Effective d^8 configurations occur in mono- and disubstituted iron carbonyls such as $Fe(CO)_4(C_2H_4)$ and $Fe(CO)_3(C_4H_4)$ or $Fe(CO)_3(C_4H_6)$, as well as in monosubstituted cobalt carbonyls $LCo(CO)_4$ (L = H, SiH_3, GeH_3). The latter compounds bear a clear resemblance in their P.E. spectra to the trigonal symmetry of $Fe(CO)_5$, exhibiting two d ionization bands of equal intensity at 8.8–8.9 and 9.9 eV, as corresponding to the $C_{3v}(1 e)^4 (2 e)^4$ configuration *(41)*; σ_{Co-H} is ionized at 11.50 eV. No σ_{Co-Si} or σ_{Co-Ge} ionization is observed as a separate band, but σ_{Si-H} and σ_{Ge-H} give rise to a band at 11.90 eV *(41)*. Less clear, and sometimes conflicting assignments are reported for substituted iron carbonyls: In $Fe(CO)_4(C_2H_4)$ the d-ionization region shows, in contrast to the two sharply defined peaks of $Fe(CO)_5$, a smeared out broad band where four or five maxima can be distinguished, possibly corresponding to the four filled d orbitals (23), while two subsequent bands at 10.5 and 12.3 eV closely match in energy the first two ionization bands of ethylene and are likewise assigned as $\pi(C=C)$ and $\sigma(CH)$ (23). In the butadiene derivative $Fe(CO)_3(C_4H_6)$ a strong mixing of metal orbitals and ligand-originated orbitals seems to occur; ionization of nine orbitals (four metal-d, two ligand π, and three ligand σ) were identified in the region of I.P. < 14 eV, that is, before CO ionizations, in a sequence of four composite bands to which orbital multiplicity 4, 2, 1, and 2 were assigned (14). $Fe(CO)_3(C_4H_4)$ has a similar and simpler spectrum, the lower I.P. region containing two broad bands. The first is peaked at 8.45 eV and is thought to contain the ionizations from three of the predominantly metal-d orbitals; the second, at 9.21 eV, contains two orbitals deriving from ligand π, while a fourth metal orbital lies probably at lower orbital energy (17). Again quite similar, both in the observed spectral patterns and in the proposed interpretation, is the spectrum of the trimethylenemethane complex $Fe(CO)_3[C(CH_2)_3]$ $(18, 46)$. Several other iron tricarbonyl diolefine complexes were investigated earlier by *Dewar et al.* (47) under low-resolution conditions; however, only the lowest adiabatic I.P.s were recorded, and the results are difficult to compare with those from later works. For further diolefin complexes, see (118).

The d^{10} tetrahedral structure of $Ni(CO)_4$ is also present in the mixed, isoelectronic nitrosyl carbonyls $Co(NO)(CO)_3$ and $Fe(NO)_2(CO)_2$, where NO behaves effectively as a three-electron ligand, and the d-ionization patterns include, as in the former case, two bands in intensity ratio $3:2$ (the first one being split in the Fe compound where low-symmetry effects are more apparent), almost unchanged in energy from $Ni(CO)_4$ $(11, 20)$. Theoretical ab initio calculations support this assignment, although one has to resort to ΔSCF calculations since, as it often occurs with metal compounds, application of Koopmans' theorem fails because of larger relaxation effects on the ionization of metal-centered orbitals $(7-12)$; the resulting picture of the electronic structure implies a larger negative charge on NO than on CO ligands, i.e., NO is a better acceptor than CO, and the highest filled orbitals have therefore significant NO contributions, in contrast with the nearly pure metal-d character of the highest orbitals, t_2 and e, in $Ni(CO)_4$ (11).

Sandwich Metal Compounds (Table 2). Several metal complexes containing the π-cyclopentadienyl ligand and its substituted derivatives have been studied by photoelectron spectroscopy. The P.E. spectrum of free cyclopentadiene (48) shows two bands of approximately equal height at 8.57 and 10.72 eV, due to ionization of the antisymmetric and symmetric combinations of the two π bonding orbitals, followed by a series of bands mainly associated with σ ionizations. The situation is somewhat different in the π^6 system cyclopentadienide ion $C_5H_5^-$, where the five carbon $2p\pi$ atomic orbitals give rise to three occupied π M.O.s with the same relative ordering as in the π-isoelectronic system benzene, $e_1''(\pi)$ and $a_2''(\pi)$ (in most metallocenes, ring rotation probably occurs rather easily, but a D_{5d} point symmetry is usually adequate for the analysis of P.E. spectra).

In the P.E. spectra of metal bis-cyclopentadienides (Fig. 6), starting with d^0 systems like Mg cp$_2$ (*28, 49*), the highest occupied π level of the ligands becomes a set of two symmetry-split levels $e_{1g} + e_{1u}$ (in a D_{5d} point group) either unresolved or separated by no more than ~ 1 eV in the corresponding P.E. ionization bands. Both bands are nearly equal in width, and a nonbonding character has been attributed to them, in agreement with the assumed ionic nature of $Mg(\pi\text{-}C_5H_5)_2$. The complex profiles of such bands have been attributed to vibronic and Jahn-Teller effects (*49*). Two subsequent systems of broad bands centered at ~ 12.5 and 17.0 eV are assumed to be due predominantly to σ electrons of the ligand framework, whose ionization probably masks the ionization of the remaining π orbitals. Quite similar is the P.E. spectrum of Mg(Mecp)$_2$ (*49*). More complex are the spectra of the pseudotetrahedral dihalides of titanocene, zirconocene, and hafnocene (*50*), all of which show a system

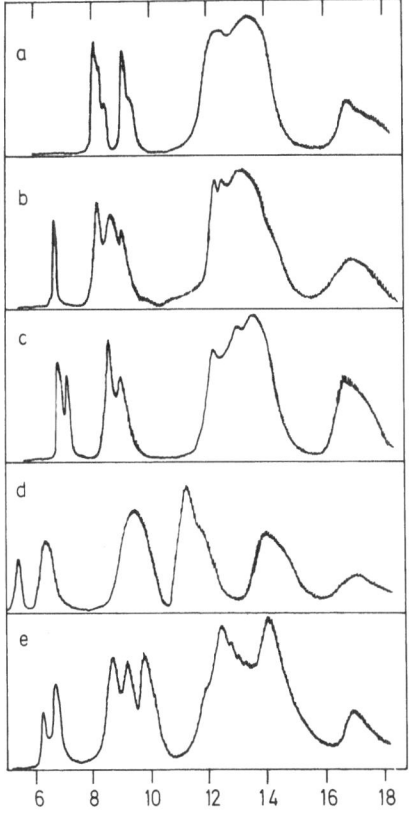

Fig. 6. He(I) P.E. spectra of sandwich complexes: (a) Mgcp$_2$ (*49*); (b) Vcp$_2$ (*52*); (c) Fecp$_2$ (*49*); (d) Crbz$_2$ (*16, 57*); (e) Mn(cp)(bz) (*57*)

Table 2

Compounds	Formal metal electronic configuration	Ionization energies (eV)[a]						References[b]
		"d"[c]	"e₁g"[c]	"e₁u"[f]	Other	Ligands σ, π	Ligands σ	
bz			9.25			11.49 12.2 12.6 13.2 13.8 14.8	16.8	6
cpH			8.57[g]		10.72[h]	12.4 12.5 13.3 14.5	16.4	49
mecpH			8.28[g]		10.45[h]	12.2 12.5 13.5	16.9	49
Mgcp₂	d^0		8.11	8.23 8.44	9.26	11.7 12.4 13.0	16.65	28, 49
Mg(mecp)₂	d^0		7.78	7.90 8.10	8.86		15.9	49
TiCl₂cp₂	d^0		8.46	8.60 (8.87) 9.07	(9.95) 10.24 10.65 11.12[i]	13.1 13.8	17.1	50
ZrCl₂cp₂	d^0			9.08	19.84 10.45 11.12 11.33[i]	13.1 13.8	17.1	50
HfCl₂cp₂	d^0			9.30	10.00 10.60 11.32 11.60[i]	13.2 13.7	17.8	50
Tacp₂H₃	d^0			9.6	[8.1] 9.5[k] 10.6[j]	12.7 13.5	16.9	51
Mocp₂H₂	d^2	6.4		[8.9]	8.7 9.5[k] 11.3[j]	12.5 13.5	16.8	51
Mo(CH₃)₂cp₂	d^2	6.1			9.61 11.3[j]	12.6	17.1	51
Wcp₂H₂	d^2	6.4		[8.9]	8.7	12.6 13.7	17.0	51
W(CH₃)₂cp₂	d^2	6.0		8.8	9.0[k] 9.61 11.3[j]	12.6 13.3	17.1	51
Vcp₂	d^3	6.78	8.40	8.65 8.79 9.02	8.9	(11.85) 12.37 13.10 13.35	16.8	52
V(mecp)₂	d^3	6.60	(8.03)	8.39 8.73 9.02		11.79 12.37 13.10 13.91	16.02 16.87	52
Crcp₂	d^4	[5.71] 7.04 7.30 (7.58)[m]	8.55	8.73 (8.87) 9.25 (9.57)	9.26	(11.89) 12.25 (12.82) 13.44	16.8	28, 52, 53, 54
Cr(mecp)₂	d^4	[5.53] 6.85 7.10 7.41[n]	8.32	9.02		11.79 12.39 12.94 13.93	15.96 16.81	52, 54
Ticotcp	d^4	6.83	8.71	9.1 10.2	10.2	11.5 12.3 12.8 14.6	16.2 16.6	55, 56
Zrcotcp	d^4	6.94	8.89	9.26 10.2	10.2	11.7 12.5 13.2 14.5	16.0 16.6	55
Recp₂H	d^4	6.4[n]		8.8	9.2 9.9[k]	12.6 13.4	17.0	51
Mncp₂	d^5	6.91[o] 10.10[f] 7.00[o]	8.76	8.42 8.75		12.37 13.18	16.71	52, 53
Mn(mecp)₂	d^5	6.58 7.15 7.36	8.42	8.75		12.47 13.09	15.98 16.86	52
Vmes₂	d^5	6.01				10.9 11.4 12.0 13.0	15.6 17.8	56
Crbzcp	d^5	5.33[t] 5.61[u] 6.08[v]	8.76	9.17 9.68	10.23	11.5 12.2 13.8	17.2	57
Vcctcp	d^5	6.42[n] 6.77 7.28[y]	[8.66]	(8.99)[z] 9.13[z] 10.2(a')		11.5 12.2 12.8 14.6	16.6	56
Nbcctcp	d^5	5.98[n] 7.11 7.50[y]	[8.78]	9.13[z] 10.4(a')		11.7 12.2 12.5 13.6 14.5	16.3 17.2	55
Ticotcp	d^5	5.67[n] 7.62[z]	[8.63]	(8.93)[z] 10.51(b')		11.5 12.6 14.8 15.1	16.6	55
Crbz₂	d^6	5.4[f] 6.4[s]		9.6		11.5 13.8	16.6	16, 57
Crtol₂	d^6	5.24[f] 6.19[s]		9.53		11.6 13.8 14.7	16.2	57
Crmes₂	d^6	5.01[f] 5.88[s]	8.90			10.9 11.6 12.1 13.4	17.8	56
Mobz₂	d^6	5.52[f] 6.59[s]		9.47 10.15		11.7 12.1 14.2	16.6	56
Motol₂	d^6	5.32[f] 6.33[s]		9.05 9.75		11.2 11.8 13.8 14.7	16.2	56
Momes₂	d^6	5.13[f] 6.03[s]		8.63 9.33		10.9 11.5 12.0 13.0		56
Mnbzcp	d^6	6.36[f] 6.72[s]	8.75	9.25 9.79		11.4 12.2 14.1	16.8	57
Crcetcp	d^6	5.59[f] 7.19[s]	[8.69]	(9.00)[z] 10.4(a')		11.5 12.1 12.7	16.5	56
Mocetcp	d^6	5.87[f] 7.55[s]	[8.93]	(9.28)[z] 10.4(a')		11.7 12.3 12.9 13.6 14.8	16.6	56
Fecp₂	d^6	6.88[s] 7.23[s]	9.14	9.39	8.72 (8.87)	12.3 13.0 13.46	16.5	28, 49, 53, 58, 85
Fe(mecp)₂	d^6	6.72[f] 7.06[s]		9.17	8.53 (8.73)	11.81 12.58 13.23	16.02 16.89	49
Fe(Clcp)₂	d^6	7.03[s] 7.37[s]	9.17	8.71 9.09	10.98 11.44[f']	(12.4) 13.58	16.63	49
Rucp₂	d^6	7.45[s] (7.63)[f]	9.93	8.51 (8.80)		11.8 12.3 13.4	16.75	49
Ru(mecp)₂	d^6	7.25		9.76 (10.23) 8.24 (8.40)		11.72 12.33 13.26	15.95 17.69	49
Os(mecp)₂	d^6	6.93		9.90 8.26 8.68		11.42 11.81 13.51	15.98 16.75	49
Cocp₂	d^7	5.56 7.18 7.63 8.01	8.66	(8.94) (9.31)	9.88	12.24 13.99	16.57	28, 52
Co(mecp)₂	d^7	5.37 6.97 7.43 7.80	8.40	(8.64) (9.03)	9.59	11.72 12.45 12.98	15.94 16.89	52
Nicp₂	d^8	6.51[q] 6.36[q]	[8.43]	9.22	10.33[h'] 10.00[h']	12.25 (12.84) 13.27	16.59	28, 52, 53
Ni(mecp)₂	d^8	6.79[s] 7.91[i]	[(8.30)]	9.05	10.65	(11.77) (12.49) 13.05 (14.00)	16.0 16.9	52
Thcot₂	d^0f^0	7.85[f]	9.90	10.14	10.56	11.48 12.32 14.12 14.65	16.17 16.74 17.91	61
Ucot₂	d^0f^2	6.20[j] 6.90[s]	9.95	10.28		11.50 12.37 14.09 14.67	16.10 16.73 17.85	61

of four bands. CNDO calculations suggest that the highest filled M.O. is essentially localized in the cyclopentadienyl rings, while those localized on the halogen atoms are expected at lower orbital energies, so the bands at ~ 9.0 and 11.0 eV are assigned to ring-π ionizations, and the broad system centered at ~ 14 eV, to halogen π pairs, although a certain degree of intermixing between cp and halide orbitals is to be assumed in all cases.

The P.E. spectrum of cp_2TaH_3 (51) has no bands attributable to d-metal electrons, in accordance with an effective d^0 configuration. There are four distinct bands between 8 and 11 eV, assigned altogether to the ionizations of the highest π bonds of the ligand rings and of the metal-hydrogen bonding orbitals. *Green et al.* observed d-bands in the spectra of some related cyclopentadienyl metal hydrides (51) such as cp_2MH_2 (M = Mo, W): a band at 6.4 eV in both complexes is clearly related to a lone pair mainly localized on the metal atom, while ionizations from metal-cyclopentadienide and metal-hydrogen orbitals merge into a broad band centered at about 9 eV.

The P.E. spectra of open shell metallocene species were extensively investigated by *Evans* and co-workers (52) and can be divided into three ionization regions. The highest I.E. region is assigned to strongly bonding σ M.O.s, the intermediate region $(\sim 11.5-15.0 \, eV)$ to two π-bonding ligand orbitals plus less bonding σ orbitals of the metal sandwich system, while the lower I.E. region, below 11 eV, shows a more complex structure and involves ionizations of essentially metal d electrons, just mixed to some extent into partly ligand-delocalized π-orbitals. Assignments in this region are supported by ligand-field computations and by comparison with absorption spectroscopic data (42). In Vcp_2 $(V(Mecp)_2)$ the three d electrons give rise to one band at 6.78 (6.60) eV including both ionic states $^3A_{2g}$ and $^3E_{2g}$ (d^2 in D_{5d} symmetry); the band system at about 9 eV has a diffuse structure as a result of coupling with the d-open shell.

The d^4 system $Crcp_2$ was extensively investigated by several authors $(28, 53, 54)$, some ambiguities persisting in the interpretation of the region of lower I.E., where four bands are observed between ~ 5 and 8 eV. *Green et al.* (51) propose, on the ground of energetic and intensity considerations, the I.E. ordering $^4A_{2g} < ^2E_{1g} < ^2A_{1g} < ^2A_{2g}$, while the fifth expected ionic state should be masked by the first ligand ionization. *Cox et al.* (54) agree on such assignment, except for the identification of

---

a Shoulders in parentheses.
b Numerical data are taken from the references in italics.
c e_{1g}-type orbital deriving from the highest "π" orbital of the aromatic ligand.
f e_{1u}-type orbital deriving from the highest "π" orbital of the aromatic ligand.
g Antisymmetric combination of the π orbitals of the conjugated double bonds.
h Symmetric combination of the π orbitals of the conjugated double bonds.
i Halogen σ, π orbitals.
j M–H σ bonding orbitals.
k π(cp) + σ M–H.
l CH_3 orbitals.
m $^4A_{2g} + ^2E_{1g} + ^2A_{1g} + ^2A_{2g}$.
n a_1-type sublevel.
o b_1-type sublevel.
p Very weak band.
q e_{1g}-type sublevel.
r a_{1g}-type sublevel.

s e_{2g}-type sublevel.
t $^3E_{2g}$.
u $^1A_{1g}$.
v $^1E_{2g}$.
w $^1A_1 + ^3E_2$.
x $^1E_2$.
y e_2-type sublevel.
z e_1(cp).
a' e_1(cet).
b' e_1(cot).
f' Halogen π orbitals.
g' a_{1g}-type sublevel + e_{2g}-type sublevel.
h' $e_{2g}(d) + e_{1g}(\pi) + e_{1u}(\pi)$.
i' e_{2u}-type sublevel.
j' Metal "f" orbital.

both $^4A_{2g}$ and $^2E_{1g}$ as contributing to the first band. *Rabalais et al.* (*53*) prefer the ordering $^2E_{2g} < {}^4A_{2g} < {}^2E_{1g} < {}^2A_{1g} \sim {}^2A_{2g}$, but this assignment requires the $a_{1g}(d)$ cross section to be more than twice that of $e_{2g}(d)$; anyway, the spectrum is not compatible with any simple crystal field interpretation. The P.E. spectrum of Cr(Mecp)$_2$ (*52,54*) is fully analogous.

Much simpler is the spectrum of other d^4 species such as Ticp(cet) (*55,56*) and Zrcp(cet) (*55*), having a ground-state diamagnetic configuration $(e_2)^4$; in accordance with such configuration, the P.E. spectra have a band at 6.83 (6.94 in the Zr species) eV assigned to the ionization of the e orbitals. The bands at 8.7 and 10.2 eV in Ticp(cet) are characteristic of the cyclopentadienyl e_1 and cycloheptatrienyl e_1 ionizations, respectively.

The four d-metal electrons in cp$_2$ReH give rise to two P.E. bands (*51*) at 6.4 and 7.0 eV, while the remaining part of the spectrum is similar to that of cp$_2$TaH$_3$ discussed previously.

The d^5 manganocene and dimethylmanganocene systems present some ambiguities in the interpretation of their P.E. spectra. Thus, Mncp$_2$ and Crcp$_2$ were investigated by the same authors (*52, 53*), who are in disagreement on the assignment of the low I.E. region. For Mncp$_2$, the high-spin groundstate configuration $(a_{1g})^1$-$(e_{2g})^2(e_{1g})^2$ appears well established in the solid state and in solution, but the situation is less certain in vapor phase. *Evans et al.* (*52*) propose an assignment assuming a high-spin configuration: the band with main maximum at 6.91 is attributed to the ionization of the $e_{1g}(d)$ electrons leading to the $^5E_{1g}$ ion state, while the weaker band system in the region 10–10.5 eV is assigned to the $a_{1g}(d)$ ($\rightarrow {}^5A_{1g}$) and e_{2g} ($\rightarrow {}^5E_{2g}$) ionizations. This interpretation is in agreement with crystal field considerations. The second band, broad and centered at 8.76 eV, is related to the various ligand π ionization processes. *Rabalais et al.* (*53*), on the contrary, assume the low-spin configuration $e_{2g}^4 a_{1g}^1 ({}^2A_{1g})$ in the groundstate, giving rise upon ionization to the ionic states $^1A_{1g}$, $^3E_{2g}$, and $^1E_{2g}$. The authors relate these ionic states to the first band, which appears under high resolution to be composed of three components. The spectrum of Mn(Mecp)$_2$ is even more complicated, and *Evans et al.* (*52*) suggest an equilibrium in the gas phase between high and low-spin states. There are no ambiguities in the P.E. spectra of the isoelectronic Ti(cp)(cot) and V(cp)(cet) (*56*) and Nb(cp)(cet) (*55*), which have only one unpaired electron. The first three bands in V(cp)(cet) and Nb(cp)(cet) are related to the ionic states $^1A_1$, $^3E_2$ and $^1E_2$ in order of increasing I.E., while in Ti(cp)(cot) the $^3E_2$ and $^1E_2$ states coalesce to a single, broader band.

The spectrum of Vmes$_2$ (*56*) having $^2A_{1g}$ ground-state is similar to the ones just described, except that the first two bands are interchanged (as is evident from the intensity ratios); the assignment is therefore to $^3E_{2g} < {}^1A_{1g} < {}^1E_{2g}$ in order of increasing I.E.

Several mixed π-benzene-π-cyclopentadienyl complexes have been investigated in UPS in order to compare the properties of these groups as ligands. The d^5 system Cr(cp)(bz) was investigated by *Evans et al.* (*57*) together with the d^6 complexes Cr(bz)$_2$, Cr(tol)$_2$, and Mn(cp)(bz). The spectrum of Cr(cp)(bz) shows in the region

8–10 eV three or four bands although the ligands have only two bands in the same energy region, originating from the two e_1 lone pair orbitals, one from benzene, and one from the cyclopentadienyl ligand. The removal of degeneracy of the e_1 orbitals can be explained by Jahn-Teller effect or by assuming an effective symmetry lower than $C_{\infty v}$. The I.E. region above 10 eV has bands characteristic of both the π-benzene and the π-cyclopentadienyl ligands. The assignment of the two bands at 6.20 and 7.15 eV is further complicated by the fact that the actual ground-state configuration is unknown [either $a_1{}^2 e_2{}^3 (^2E_2)$ or $a_1{}^1 e_2{}^4 (^2A_1)$]. If the ground-state is $^2A_1$, the ionizations of d electrons lead to the ionic states $^1A_1$, $^3E_2$ and $^1E_2$, so the first band should be related to $^1A_1$ and $^3E_2$, and the second, less intense one, to $^1E_2$. If the ground-state is, however, $^2E_2$, the arising ionic states are $^3A_2$, $^1E_2$, and $^1A_1$ related to the first band, and $^3E_2$, $^1E_2$ related to the second band. *Evans et al.* (*57*) prefer the former assignment, which is supported by magnetic susceptibility and ESR measurements.

Bisbenzene chromium and bistoluene chromium display, in the lower I.E. region of their P.E. spectra, two bands in the region below 8 eV, easily attributed to ionizations from the fully occupied e_{2g} and a_{1g} d-orbitals. Above 8 eV there is strict similarity to the P.E. spectra of the free arene ligand and the assignment is obvious, so ionization of the arene e_{1u} and e_{1g} orbitals occur between 8.5 and 10 eV. Crbz$_2$ was investigated also by *Guest et al.* (*16*), who substantially agree with *Evans et al.* (*57*) and support the assignment by theoretical considerations as well as by comparison with He(II) spectra, although intensity considerations suggest smaller d-participation in the second band.

The P.E. spectra of Cr(mes)$_2$, Mo(bz)$_2$, Mo(tol)$_2$, and Mo(mes)$_2$ (*56*) are similar to Cr(bz)$_2$ in the region of d-ionizations ($<$ 8 eV), with no relevant differences. Also the isoelectronic Mn(cp)(bz) (*57*) displays a similar P.E. spectrum below $\sim$ 8 eV, with two bands at slightly higher I.E. than the previously mentioned compounds, assigned to a_{1g} and e_{2g} d-type orbitals; the region beyond 8 eV is again analogous, for example, to Cr(cp)(bz) (*57*).

Cr(cp)(cet) (*56*) and Mo(cp)(cet) (*55*) show analogously their a_1 and e_2 bands at about 6.0 and 7.5 eV.

The P.E. spectrum of ferrocene was investigated by several authors (*28, 49, 53, 58, 59*). *Evans et al.* (*49*) extended their investigation to other d^6 systems Fe(Mecp)$_2$, Fe(Clcp)$_2$, Ru(cp)$_2$, Ru(Mecp)$_2$, and Os(Mecp)$_2$. Ferrocene displays in the region of lower I.P. two bands, at 6.88 and 7.23 eV, with approximate intensity ratio 2:1. Therefore, the order of energy levels of a_{1g} and e_{2g} is reversed with respect to the other d^6 sandwich complexes previously described. The remaining part of the spectrum contains bands related to orbitals of the cyclopentadienide rings, and the assignment does not pose any problem of particular interest. Very similar are the P.E. spectra of Fe(Mecp)$_2$ and Fe(Clcp)$_2$ (*49*), apart from a shift in I.E.s due to the inductive effect of the methyl or Cl substituents. In the spectrum of the latter compound two bands are present, at $\sim$ 11.0 and 11.4 eV, which have no counterpart in the spectrum of ferrocene and are therefore assigned to ionization of essentially chlorine $3p_\pi$ lone pair electrons.

The P.E. spectrum of ruthenocene and 1,1'-dimethylruthenocene (49) are grossly similar to those of the corresponding iron compounds, but there are some significant differences in the region of low I.E. Thus, the two P.E. bands corresponding to the a_1 and e_2 d-type orbitals overlap in the spectra of the ruthenium complexes, and the I.E. of the single band is by ~ 0.5 eV higher than the average of the corresponding bands in Fe(cp)$_2$, indicating a greater "covalent" character in the ruthenium compounds.

Os(Mecp)$_2$ shows in its P.E. spectrum three distinct bands below 8 eV, due to spin-orbit splitting of the $^2E_2(5/2)$ and $(3/2)$ ionic states.

The d^7 cobaltocene and dimethylcobaltocene molecules have the ground-state configuration $a_{1g}^2 e_{2g}^4 e_{1g}^1 (^2E_{2g})$. Their P.E. spectra, investigated by Evans et al. (52), show four bands in the d-region. The first band is assigned to ionization of the solitary $e_{1g}(d)$ electron to form the $^1A_{1g}$ ionic state; the second band, corresponding to the almost coincident $a_{1g}^2 e_{2g}^3 e_{1g}^1 (^3E_{1g})$ and $a_{1g}^2 e_{2g}^3 e_{1g}^1 (^3E_{2g})$ ionizations, completes the picture of d ionizations. The origin of the third and fourth band is less clear; the authors (52) consider up to four possible alternative assignments involving the $^1E_{2g}(a_{1g}^2 e_{2g}^3 e_{1g}^1)$, $^1E_{1g}(a_{1g}^1 e_{2g}^4 e_{1g}^1)$ and $^3E_{1g}(a_{1g}^1 e_{2g}^4 e_{1g}^1)$ ionizations.

The ground state of the open-shell d^8 molecule nickelocene is $^3A_{2g}(a_{1g}^2 e_{2g}^4 e_{1g}^2)$, and its P.E. spectrum, investigated by (42) and (53), is complicated by the overlap of metal d and ligand π ionizations. The solitary, well resolved band at ~ 6 eV must clearly be assigned to ionization of an $e_{1g}(d)$ electron leading to the ionic state $(a_{1g}^2 e_{2g}^4 e_{1g}^1)\, ^2E_{1g}$. In the diffuse band system between ~ 0.8 and 10.5 eV, Evans et al. (52) identify four distinct peaks, Rabalais et al. (53) only two, but they all agree in assigning these bands to intermixing metal d and $\pi\,(e_{1g} + e_{1u})$ ligand ionizations. The spectrum of Ni(Mecp)$_2$ is very similar. Evans et al. (60) investigated the d^{10} systems Ni(cp)(NO) and Ni(Mecp)(NO) with the ground-state configuration $e_1^4 e_2^4 a_1^2 (C_{5v})$. Of the four bands observed between ~ 8 and 11 eV, the second and third ones are related to $a_1(d)$ and $e_2(d)$ ionizations, while the first and fourth bands are both related to e_1 ionizations, although d and ligand π ionizations cannot be sorted out singly. In the P.E. spectrum of Ni(Mecp)(NO) the first two ionization bands are apparently interchanged, in accordance with the ligand π character of the involved M.O.s, which is obviously very sensitive to the presence of methyl substituents.

There are only two examples of sandwich actinide complexes investigated by P.E. spectra: thorocene and uranocene (61). The ground-state electronic configuration is $a_{1g}^2 a_{2u}^2 e_{1u}^4 e_{1g}^4 e_{2u}^4 e_{2g}^4 f^n$ for both Th(cot)$_2$ and U(cot)$_2$, with $n = 0$ for the thorium and $n = 2$ for the uranium compound. The P.E. spectrum is very similar for both compounds, apart from the lack of the first band in thorocene, which is instead present in uranocene, and clearly due to ionization of f electrons. The next two bands, at ~ 7 and 8 eV, in both compounds are assigned to ionization of the e_{2g} and e_{2u} ring orbitals. The P.E. spectra of these cyclooctatetraene complexes show, on the whole, striking resemblances to those of other transition metal sandwich complexes with π-ligands like benzene and cyclopentadienyl, C_6H_6 and $C_5H_5^-$.

Table 3

Compounds	Formal metal electronic configuration	Ionization energies (eV)[a]					References[b]
		d''	e_{1g}[c], e_{1u}[f]	Other	Ligands π, σ	Ligands σ,	
$Nb(C_3H_5)cp_2$	d^2	5.7	|8.6 9.3 9.2|[g]	8.0[h]	13.0 15.0 16.9		51
$Mo(CO)_2cp_2$	d^4	5.9[i] 6.8[j]	|8.8 9.6|[k]		12.6 13.6	16.8	51
$Mo(C_2H_4)cp_2$	d^4	6.0[i] 6.9[j]	|8.8 9.3 9.2|[k]	11.3[h]	12.3 13.2 13.0	17.0	51
$W(C_2H_4)cp_2$	d^4	6.0[i] 7.1[j]	|9.0 9.5|[k]	11.3[h]	12.5 13.0	16.7	51
$W(C_3H_6)cp_2$	d^4	5.9[i] 7.0[j]	|8.9 9.3 9.5|[k]	11.0[h]	13.2	16.9	51
$Mo(CH_3)(CO)_3cp$	d^4	7.78[l]	9.7 10.0	9.07[h] (12.2)[m]			62
$W(CH_3)(CO)_3cp$	d^4	|(7.60) 7.77|[l]	9.92 10.2	9.26[h] (12.3)[m]	13.8	17.3	62
$Mn(CO)_3cp$	d^6	8.05[n] 8.40[i]	9.90 10.29		13.3 13.8	17.3	43
$Mn(CO)_2cpN_2$	d^6	7.54 7.89 8.07	9.78 10.17				65
$Mn(CO)_2cpNH_3$	d^6	6.63 6.99 7.36	9.15		13–15		65
$Mnchd(CO)_3$	d^6	8.06	8.59[o] 10.27[p]	[~11–12][q]		~14	64
$Mnced(CO)_3$	d^6	7.86 8.10	8.67[o] 9.97[p]	[~11–12][q]		~14	64
$Mncet(CO)_3$	d^6	7.78	8.33[o] 10.33[p]			~14	64
$Re(CO)_3cp$	d^6	8.13 8.52 8.76	10.18 10.59				43
$FeCl(CO)_2cp$	d^6	|8.00 8.38|[r] 8.99[s]	|10.5–11.1|[t]	10.17[u]			43, 63
$FeBr(CO)_2cp$	d^6	|7.93 8.30|[r] 8.99[s]	|10.4–10.8|[t]	9.78[u]			43, 63
$Fe(CO)_2cpI$	d^6	|7.77 8.17|[r] 8.73[s]	10.40 10.76	9.37[u] 10.3[v]			43, 63
$Fe(CH_3)(CO)_2cp$	d^6	|7.79 8.03|[n] 8.61[s]	9.90 10.26	9.23[h]			43, 62, 63
$Ru(CH_3)(CO)_2cp$	d^6	8.13 8.29 8.96	9.98 10.51	9.48[h]	>13		62
$Fe(CNCH_2)(CO)_2cp$	d^6	8.29 8.90 9.48	10.25	11.14[h] [11.89 [12.27][w]	13.2 14.0 14.6	17.7	62
$Crbz(CO)_3$	d^6	7.42	10.70		12.70 15.01 16.03 16.54	17.86	16
${Fe(CO)cp}_4$	d^6	6.45 6.87 8.58	8.9 9.15		11.30 11.90 12.30	15.4	63

[a] Shoulders in parentheses.
[b] Numerical data are taken from the references in italics.
[c] e_{1g}-type orbital deriving from the highest "π" orbital of the aromatic ligand.
[f] e_{1u}-type orbital deriving from the highest "π" orbital of the aromatic ligand.
[g] $1a' + 2a'' + 2a' + 3a'$.
[h] σ M–C.
[i] a_1-type sublevel.

[j] b_1-type sublevel.
[k] $a_1 + b_2 + b_1 + a_2$.
[l] a'-type sublevel + a''-type sublevel.
[m] CH_3 orbital.
[n] e-type sublevel.
[o] $a'(e')$.
[p] $a''(e'')$.

[q] σ C–H orbitals.
[r] $e(d + p$ halogen).
[s] b_2-type sublevel.
[t] cp + σ-bonding M-halogen.
[u] e(π-bonding M-halogen).
[v] σ-bonding M-halogen.
[w] π(CN).

Mixed Carbonyl-Cyclopentadienides and Carbonyl-Arene Metal Complexes (Table 3). Extensive P.E. investigations have been carried out in this field, and their interpretation confirms the availability of three orbitals for further coordinative bonding in each M(cp) unit *(51)*. *Green et al. (51)* reported data on several cyclo-pentadienyl or olefine mixed carbonyls (Fig. 7). The nature of bonding in olefin-transition metal complexes is formally similar to that of carbon monoxide-transition metal complexes, so we shall discuss both classes of compounds altogether. The spectrum of the d^2 system $Nb(cp)_2(C_3H_5)$ *(51)* shows a narrow band at 5.7 eV, easily assigned to the niobium d lone pair ($4a'$ in the C_s point group), followed by a broader band at 8 eV, related to the σ bonding metal-π-allyl orbital $3a''$. The two bands at 8.6 and 9.2 eV can be assigned to the metal-cyclopentadienyl orbitals $1a'' + 2a'' + 2a' + 3a'$. Very similar is the P.E. spectrum of $Mo(cp)_2(C_2H_4)$ *(51)*: The band at 5.9 eV corresponds to the band at 5.7 eV in $Nb(cp)_2(C_3H_5)$, while the band at 6.9 eV may be related to the band at 8 eV in the latter compound. However, in the molybdenum compound the second ionization is more closely concerned with the metal d_{xy} orbital. The bonding σ_{CH}, σ_{CC}, and other cyclopentadienyl orbitals give rise to the broad envelope of bands beyond 11.5 eV in both compounds. The peak at 11.3 eV in $Mo(cp)_2(C_2H_4)$ is assigned to the $1a_1$ orbital mainly formed by the olefin π orbital. This band is not observed in the P.E. spectrum of $Nb(cp)_2(C_3H_5)$, probably masked by other ionizations comprised in the broad envelope beyond 11.5 eV.

The P.E. spectra of $W(cp)_2(C_2H_4)$ and $W(cp)_2(C_3H_6)$ are very similar to $Mo(cp)_2(C_2H_4)$, and the assignment is again analogous *(51)*. Still on the same line is

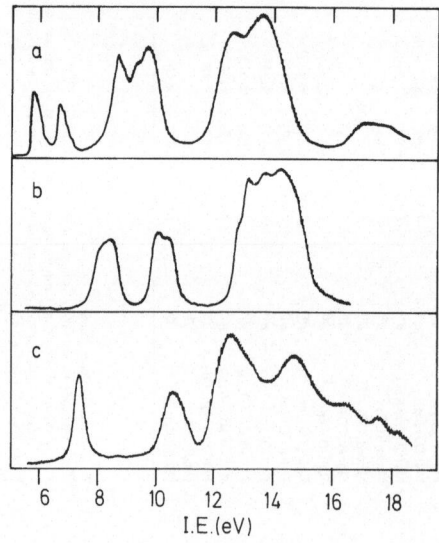

Fig. 7. He(I) P.E. spectra of mixed carbonyl-aromatic complexes: (a) $Mo(CO)cp_2$ *(51)*; (b) $Mn(CO)_3cp$ *(43)*; (c) $Cr(CO)_3bz$ *(16)*

also the P.E. spectrum of $Mo(cp)_2(CO)$ (51), apart from lack of the olefin band around 11 eV. The ionizations of CO orbitals fall beyond 14 eV and are therefore masked under the amorphous σ_{CH} and σ_{CC} ionization bands.

In the P.E. spectra of $M(cp)(CO)_3(CH_3)$ (M = Mo, W) (62) only one band, falling at ~ 7.8 eV, is observed corresponding to both metal d orbitals $6a'$ and $3a''$; the band at ~ 9 eV is assigned to the $5a'$ orbital (metal-carbon σ bond), while the two overlapping bands around 10 eV are assigned to the essentially cyclopentadienyl orbitals $4a'$ and $2a''$. The same authors (62) investigated the d^6 systems $M(cp)(CO)_2(CH_3)$ (M = Fe, Ru). Considering the $(CH_3)(CO)_2$ ligand system as exerting the same ligand function as a cyclopentadienyl group, the spectra of these compounds can be compared to those of the corresponding $M(cp)_2$. The bands at 7.78 and 8.53 in $Fe(cp)$-$(CO)_2(CH_3)$ resemble the e_{2g} and a_{1g} bands of ferrocene, apart from a shift to higher I.E. by $1-1.3$ eV, and are thus assigned to ionizations from mainly d-metal orbitals: The first one should be related to both the $6a'$ and $3a''$ orbitals, the second one to $5a'$. Furthermore σ metal-carbon is related to the band at 9.15 eV ($4a'$), and the $3a' + 2a''$ ionizations of the cp ligand to the broad band centered at 9.9 eV, the shoulder at 12.1 eV being due to methyl σ ionizations. The P.E. spectrum of $Fe(cp)$-$(CO)_2(CH_3)$ was investigated also by *Lichtenberger* and *Fenske* (43), besides $Fe(cp)(CO)_2(X)$ (X = Cl, Br, I); both the measured values and the proposed assignment are the same as in (62). The P.E. spectra of $Fe(cp)(CO)(X)$ (43) are very similar in the I.E. region below 9 eV, with two bands (the first of which contains two maxima) related to ionizations of essentially d electrons (although an alternative assignment as metal-X σ-bonding electrons cannot be ruled out). The next two bands, between 9 and 10.5 eV, are assigned to metal-halogen π-bonding orbitals, while the remaining part of the spectra is somewhat amorphous and contains many overlapping bands originated from cyclopentadienyl orbitals and from metal-ligand σ-bonding orbitals. Also *Symon* and *Waddington* (63) investigated the P.E. spectrum of $Fe(cp)(CO)_2(CH_3)$ and gave a different interpretation, implying however some ambiguities. They reported also the P.E. spectrum of $[Fe(cp)(CO)]_4$, which is again similar to that of $Fe(cp)_2$, apart from a strong shift to lower I.E. in the former compound.

$Ru(cp)(CO)_2(CH_3)$ shows an additional band in the region < 11 eV, probably because of a larger energy difference between the $6a'$ and $3a''$ d-type orbitals (62).

The P.E. spectrum of $Fe(cp)(CO)_2(CH_2CN)$, investigated by *Green* and *Jackson* (62), shows two sharp bands at 11.89 and 12.27 eV related to the highest $C \equiv N$ π orbital and to the nitrogen lone pair σ orbital respectively. The metal-carbon bonding $5a'$ orbital originates a band at 11.14 eV, with a shift of ~ 2 eV to higher I.E. with respect to the analogous band in $Fe(cp)(CO)_2(CH_3)$, while the band at 7.78 eV in the latter compound splits into two separate bands in $Fe(cp)(CO)_2(CH_2CN)$, and all three low-energy ionizations are therefore assigned to d-orbitals.

Lichtenberger and *Fenske* (43) investigated the d^6 molecules $M(cp)(CO)_3$ (M = Mn, Re). The broad band centered at ~ 8.2 eV in $Mn(cp)(CO)_3$ is attributed to orbitals of mainly d-character, while the two overlapping bands at 9.90 and 10.29 eV originate from predominantly ring orbitals. The spectrum of $Re(cp)(CO)_3$ is similar, except that three peaks are evident in the first band, the further splitting being as-cribed to spin-orbit splitting in the positive ion.

Other cyclic olefine-manganese tricarbonyl complexes were investigated by *Whitesides* and *Lichtenberger* (*64*), and include $(\eta^5\text{chd})\text{Mn(CO)}_3$, $(\eta^5\text{ced})\text{Mn(CO)}_3$, and $(\eta^5\text{cet})\text{Mn(CO)}_3$. In all three complexes, the ionizations of the six d-electrons fall at ~ 8 eV, giving rise to a broad band, while π orbitals of the cyclic ligands produce two bands, at ~ 8.5 and 10 eV.

The effect of replacing one of the carbonyl ligands through a nitrogen donor was observed by *Lichtenberger et al.* (*65*) on $\text{Mn(cp)(CO)}_2(\text{N}_2)$ and $\text{Mn(cp)(CO)}_2(\text{NH}_3)$. Major differences are evident in the ionizations associated predominantly with the central metal atom: while in Mn(cp)(CO)_3 a single band with no discernible shoulders is observed at ~ 8 eV, two distinctly separate vertical I.E.s appear in the same spectral region for $\text{Mn(cp)(CO)}_2(\text{N}_2)$, in intensity ratio approximately $1:2$; one must therefore assume a fairly larger splitting of d-levels in the latter compound. The same spectral patterns appear also in the P.E. spectrum of $\text{Mn(cp)(CO)}_2(\text{NH}_3)$. Furthermore, a shift to lower I.E. values is observed on passing from Mn(cp)(CO)_3 to the nitrogen-sub-stituted analog, in accordance with a larger electron density at the metal, probably related to the poorer π-acceptance of N ligands if compared to CO.

The only example of P.E. spectra of a mixed arene-carbonyl metal complex is that of Cr(bz)(CO)_3 reported by *Guest et al.* (*16*). In the latter compound, the two d-bands observed in the spectrum of Cr(bz)_2 merge into a single band, with maximum at 7.42 eV, containing the ionizations of all six d-electrons. A second band, at 10.70 eV, is related to a benzene orbital, while a third band, at 12.70 eV, has both carbonyl and benzene character. The assignment is supported by theoretical calcula-tions and by comparison between He(**I**) and He(**II**) spectra (*16*).

Complexes of d-Metals in Higher Oxidation States

The situation is in some way reversed with respect to the case of low-valent d-metal complexes. Metal-based orbitals have higher I.E. and ligand-based orbitals are either the first ones to be ionized or occur at I.E.s comparable to the former ones. Ligand ionization patterns, often only slightly modified from the free ligand behavior, now constitute the dominant part of the P.E. spectra at low I.E.s, whereas d-ionizations occur sometimes at lower I.E.s (particularly with early transition elements), some-times at higher I.E.s, particularly with posttransition elements, or even in highly intermixed situations, such as occur typically with d^6 to d^8 configurations.

Oxo-, Halo- and Related Species (Table 4). Much work has been done on d^0 species, and it is interesting to discuss their P.E. spectra in some detail since the basic features are also present in all other d^n species, superimposed on the d-ionization patterns.

The molecular orbitals based on the three p atomic orbitals of each donor atom transform according to the irreducible representations of the σ and π bond systems appropriate to the MX_n structure. Thus, in a trigonal molecule like monomeric $AlCl_3$ ($66, 67$) of D_{3h} symmetry, the six π (halogen) orbitals transform like a_2' (ionized at 12.01 eV) + e' (at 12.47) + e'' (12.73) + a_2'' (13.33), and the three σ bond orbitals like e' (14.04) + a_1' (15.97). Monomeric trihalides of Al, Ga, and In have been interpreted in the same scheme ($66–68$), the only remarkable additional feature being the spin-orbit splitting of the e' levels in the iodides; splitting of the bands of e'' type, which are not spin-orbit split in the double D_{3h} group, has been taken as evidence for a distorted pyramidal (C_{3v}) symmetry of GaI_3 (68) (Fig. 8). Also the P.E. spectra of M_2X_6 dimers have been reported and in part assigned (66).

In tetrahedral species, the π orbitals transform like $t_1 + t_2 + e$ [at 11.78, 12.78, and 13.23 eV in $TiCl_4$ (69)] and the σ orbitals like $t_2 + a_1$ [13.23 and 13.97 eV in $TiCl_4$ (69)]. The observed ionization bands are not always well resolved. The actual ordering of levels may change from the sequence given above for $TiCl_4$, and it has been sometimes subject to controversy. Thus, RuO_4 has bands at 12.15, 12.92, 13.93 (15.5 sh), and 16.10 eV, assigned by *Burroughs et al.* (70) to the sequence $1 t_1 < (3 t_2 + 1 e) < 2 a_1 < 2 t_2$, neglecting the shoulder at 15.5; *Diemann* and *Müller* (71) prefer $1 t_1 < 3 t_2 (12.92) < 1 e (13.93) < 2 a_1 (15.5) < 2 t_2$, and *Forster et al.* ($72$) suggest $3 t_2 < 1 t_1 < 2 a_1 \sim 1 e < 2 t_2$. The $TiCl_4$ scheme applies also to $TiBr_4$ ($69, 70$) and GeI_4 and SnI_4 (70), for which a full treatment of the additional splitting due to spin-orbit interactions was given by (70). Related schemes, deducible from the MX_4 and MO_4 ionization patterns after allowing for symmetry descent of the molecular point group, have been developed for the assignment of the P.E. spectra of $VOCl_3$ (70), CrO_2Cl_2 ($70, 73$), and MoO_2Cl_2 (70). However, the larger spectral complexity and mixed composition of M.O.s makes actual assignment more uncertain.

VCl_4 (19) represents a clear-cut example of identification of d-ionization at low I.E. values (9.41 eV), the spectral patterns in the remaining part of the P.E. spectrum being very similar in energy to those of $TiCl_4$ (69), but different in shape since the

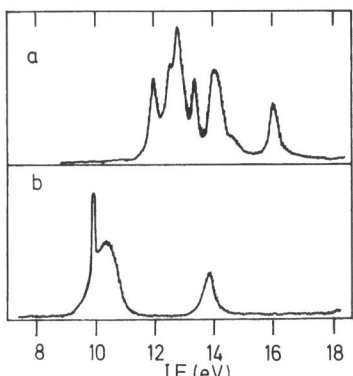

Fig. 8. He(I) P.E. spectra of gaseous metal halides: (a) $AlCl_3$ (67); (b) TlCl (83)

Table 4

Ionization energies (eV)[a]

Group 1 — trihalides (π‑halogen / σ‑halogen)

Compounds	Formal metal electronic configuration	π‑halogen					σ‑halogen				References[b]
AlCl₃	3d⁰	12.01[c]	12.47[f]	12.73[g]	13.33[h]			14.04[f]	15.97[j]		66, 67
GaCl₃	3d¹⁰ 4d⁰	11.96[c]	12.30[f]	12.44[g]	13.20[h]			13.96[f]	16.44[j]		66, 67, 68
InCl₃	4d¹⁰ 5d⁰	11.45[c,j]	12.04[f]		12.76[h]			13.22[f]	(15.02[j])	15.26[j]	67, 68
AlBr₃	3d⁰	10.91[c]	11.53[f]	11.74[g]	12.37[h]		[13.01	13.32[f]		15.23[f]	66, 67, 68
GaBr₃	3d¹⁰ 4d⁰	10.94[c]	11.25[f]	11.52[g]	12.22[h]		[12.88	13.19[f]		15.70[j]	66, 67, 68
InBr₃	4d¹⁰ 5d⁰	10.32[c,j]	[11.05	11.33[f]	~11.3[g]	11.90[h]	[12.41	12.68[f]	(14.60[j])	14.90[j]	67, 68
AlI₃	3d⁰	9.66[c]	[10.05	10.18[f]	10.46	10.56[g] 11.24[h]	[11.77	12.27[f]	14.10[j]		66, 67, 68
GaI₃	3d¹⁰ 4d⁰	9.51	9.78	9.97[f]	10.22	10.37[g] 11.00[h]	[11.51	11.99[f]	14.57[j]		66, 67, 68
InI₃	4d¹⁰ 5d⁰	9.14[j] 9.37[c]	9.65	10.36[f]	~10[g]	10.92[h]	[11.26	11.69[f]	(13.87[j])	14.12[j]	67, 68

Group 2 — dihalides (σ‑M‑halogen / π‑halogen)

Compounds	Formal metal electronic configuration	σ‑M‑halogen		π‑halogen			s²M	References[b]
SnCl₂	4d¹⁰ 5s²	12.72	10.31[k]	11.31	12.07	13.33?	15.9	87, 92
SnBr₂	4d¹⁰ 5s²	12.10	9.85[k]	10.63	11.39		15.6	87, 92
PbCl₂	5d¹⁰ 6s²	12.0	10.1[k]	10.6	11.5			112
PbBr₂	5d¹⁰ 6s²	11.63	9.81[k]	10.25	10.94	11.13	16.3	87
PbI₂	5d¹⁰ 6s²	~10.7	~8.7[k]	9.0–9.5	~10.0			112

Group 3 — oxohalides / tetrahalides (d / π‑orbitals / σ‑orbitals)

Compounds	Formal metal electronic configuration	d	π‑orbitals				σ‑orbitals			References[b]
VOCl₃	3d⁰		11.84[l]	13.65[m]	12.65	12.91	13.08[n]	13.65[n]	(13.90) 14.35[o]	70
CrO₂Cl₂	3d⁰	(11.93)	[11.85 [11.97][l]	14.30	(14.80)[m] 12.78[n]		13.24[p]	13.849		70, 73
MoO₂Cl₂	3d⁰		12.13 (12.21)[l]	14.47	14.93[m] 12.60		13.00[p]	13.769		70
OsO₄	5d⁰		12.35[l]	16.4 16.8[m]		13.54[r]	12.92[n]	13.14[s]	14.66[t]	70, 71, 72
RuO₄	4d⁰		12.15[l]	16.1[m]					13.93[t]	70, 71, 72
SnI₄	4d¹⁰ 5d⁰		9.45 10.10[l]	11.72	12.1[m]	10.29[r]	12.92[n]	9.64 9.78	10.78[s] 16.1[t]	70
GeI₄	5d¹⁰ 6d⁰	(9.42)	9.53 10.17[l]	12.05	12.40[m]	10.54[r]		(9.86)	10.89[s] 16.50[t]	70
TiCl₄	3d⁰		11.78[l]	13.23[u]				12.78[s]	13.97[t]	70
TiBr₄	3d⁰		10.63	11.68	12.00	12.31			13.04[t]	69, 70
VCl₄	3d¹	9.41	10.55 10.86[l]	(11.56) 11.68	12.00	12.31	13.08[n] 13.54[v]	(12.55)[v]	15.26[t]	19

Group 4 — silver halides (Pσ‑halogen / d)

Compounds	Formal metal electronic configuration	Pσ‑halogen	πg‑halogen	pπ‑halogen	σ‑halogen	d			References[b]
AgCl	d¹⁰	11.03		11.90	13.07?	12.78[w]	13.33[x]	15.19[y]	74
AgBr	d¹⁰	11.15		11.77		12.70[w]	14.06[x]	15.82[y]	74
AgI	d¹⁰	9.2	9.9	10.35	11.00	12.75[w]	14.10[x]	15.87[y]	74

Group 5 — group‑12 dihalides (πg‑halogen / πu‑halogen / pπ‑halogen / σu‑halogen / σg‑halogen / d)

Compounds	Formal metal electronic configuration	πg‑halogen		πu‑halogen	pπ‑halogen	σu‑halogen	σg‑halogen	d			References[b]
ZnCl₂	d¹⁰	10.89	11.87	12.39		13.07	14.10	19.0[z]	19.2[a']	19.5[b']	75, 76, 77, 13
ZnBr₂	d¹⁰	9.73		11.40		12.28	13.55	18.7[z]	18.8[a']	19.1[b']	75, 76, 77, 13
ZnI₂	d¹⁰		11.42	10.32	11.92	11.45	12.74	17.9[z]	18.58[a']	18.6[h']	75, 76, 77, 13
CdCl₂	d¹⁰	10.59		10.96	11.31	11.85	12.46	13.29	18.3[t] 19.55[f']	19.5[b']	75, 76, 13
CdBr₂	d¹⁰	9.53		10.07	10.21	11.31	11.20	12.84	19.31[i']	18.4[b']	75, 76, 13
CdI₂	d¹⁰		11.50	10.96?	12.13	10.21	11.20	12.27	19.93[f'] 20.27[l']	18.6[h']	75, 76, 13
HgCl₂	d¹⁰	11.37		10.96	12.13	12.74	11.20	13.74	17.05 16.71	17.27[i'] 18.65[r]	78, 113, 76
HgBr₂	d¹⁰	10.62		10.96	11.54	12.09	11.54	13.39	16.69 16.40	16.83[t] 18.34[j']	78, 87
HgI₂	d¹⁰	9.50		10.16	10.00	10.40	11.29	12.85	15.99 16.17	16.17[t] 17.91[k']	78

Compound	config	π-halogen, other groups	σ-orbitals	π CH₃(CF₃)	d	Ref.
Hg(CH₃)₂	d^{10}	~9.3 ~11.8		12.6 13.0 13.7|$^{k'}$	14.95 15.44|^l 16.90|$^{d'}$	78, 79
Zn(CH₃)₂	d^{10}					78
Hg(C₂H₅)₂	d^{10}				16.96|$^{i'}$	78
Hg(CH₃)Cl	d^{10}	10.88			14.68|$^{l'}$ 16.61|$^{i'}$	78
HgBr(CH₃)	d^{10}	10.16 10.43	12.70	14.1	15.79 16.24|$^{i'}$ 17.71|$^{j'}$	78
Hg(CH₃)I	d^{10}	9.25 9.68	12.52	13.9	15.68 16.111|$^{i'}$ 17.64|$^{j'}$ 17.41|$^{j'}$	78
Hg(CH₃)CN	d^{10}		12.21	13.6	15.48 15.83|$^{i'}$ 17.2$^{f'}$ 18.2|$^{l'}$ 18.8$^{f'}$	79
Hg(CN)₂	d^{10}				16.28^k 16.5^z 17.6^z 17.9^z 18.9$^{h'}$ 19.6|$^{l'}$	79
Hg(C₆H₅CH₂CH₂)Cl	d^{10}	10.73 8.65 9.33		10.73		82
Hg(CF₃)I	d^{10}	9.89 10.42	11.1$^{m'}$	12.8$^{n'}$ 15.2 19.9|$^{o'}$	16.3 16.91|$^{i'}$ 18.1|$^{i'}$	80, 114
Hg(CF₃)N₃	d^{10}	9.87 10.33	14.40$^{p'}$	12.35$^{n'}$ 15.35 15.75	16.5 16.8|$^{i'}$ 18.6|$^{i'}$	80
Hg(CF₃)NCO	d^{10}	10.83 17.85	15.15 (15.70)|$^{q'}$	12.45$^{n'}$ 17.7 20.0|$^{o'}$	16.70 17.15|$^{i'}$ 18.70|$^{i'}$	80
Hg(CF₃)NO₃	d^{10}	11.07 12.05 12.55	14.70	14.0$^{n'}$ 15.90|$^{o'}$ (20.15)|$^{o'}$	16.9 17.31|$^{i'}$ 18.8|$^{i'}$	80
Hg(SCF₃)₂	d^{10}	10.2	11.04 12.3|$^{f'}$	15.1 13.4|^s (20.0)|$^{o'}$	16.3 16.8|$^{i'}$ (18.2) (18.5)|$^{p'}$	80

		π-orbitals	σ-orbitals	π-halogen	σ-orbitals	d	Ref.
Hg(Allyl)ICl	d^{10}	9.35 10.75|$^{t'}$	11.55 12.50|$^{n'}$	13.6 14.6|$^{u'}$	15.61 ~16.2 ~17.8		81

		s²M		π-halogen	σ-orbitals	d	Ref.
InCl	$4d^{10}5s^2$	9.75		10.85	13.10		115
InBr	$4d^{10}5s^2$	9.41		10.20	12.74		115
InI	$4d^{10}5s^2$	8.82	9.10	9.78	12.21		115
TlF	$5d^{10}6s^2$	10.80		11.90	14.20	18.8|$^{i'}$ 21.0|$^{i'}$	84, 85, 86
Tl₂F₂	$5d^{10}6s^2$	19.96	10.89|$^{v'}$	11.81			84, 85
TlCl	$5d^{10}6s^2$	9.89		10.38	13.89	18.3|$^{i'}$ 20.6|$^{j'}$	83, 85, 87, 111
TlBr	$5d^{10}6s^2$	9.48		9.83	13.66	18.1|$^{i'}$ 20.4|$^{j'}$	83, 85, 116
TlI	$5d^{10}6s^2$	8.91		9.72	13.14 13.52	17.9|$^{i'}$ 20.1|$^{i'}$	83, 85, 116

a Shoulders in parentheses.
b Numerical data are taken from the references in italics.
c a'_2-type sublevel.
d e'-type sublevel.
g e''-type sublevel.
h a''_2-type sublevel.
j a'_1-type sublevel.
j adiabatic.
k b_1-type sublevel.
l $t_1(\pi)$.
m $t_2(\pi)$.
n $t_2(\sigma)+e(\pi)$.
o $a_1(\sigma)+t_2(\pi)$.
p $t_1(\sigma)+t_2(\pi)$.

q $e(\pi)+a_1(\sigma)$.
r $e(\pi)$.
s $t_2(\sigma)$.
t $a_1(\sigma)$.
u $t_2(\pi)+e(\pi)$.
v $t_2(\sigma)+t_2(\pi)+e(\pi)$.
w δ.
x π.
z $^2\Sigma_{1/2}$.
a' $^2\Pi_{3/2}+^2\Delta_{5/2}$.
b' $^2\Pi_{3/2}+^2\Delta_{3/2}$.
c' $^2\Pi_{3/2}$.
g' $^2\Delta_{5/2}$.
h' $^2\Pi_{1/2}$.

i' $^2D_{5/2}$.
j' $^2D_{3/2}$.
k' $\pi+\sigma$ orbitals.
l' $^2\Delta_{3/2}$.
m' σ Hg–I.
n' σ Hg–C.
o' CF₃-orbitals.
p' N₃ σ-orbitals.
q' NCO, CF₃ σ-orbitals.
r' σ Hg–S.
s' σ S–C.
t' Chlorine lone pairs.
u' σC–C + σC–H.
v' σ Tl–F.

presence of two open shells in the higher energy states of the VCl_4^+ ion broadens the peaks by unresolved multiplet splitting. The low d-ionization energy is in line with the expectations for early transition metal compounds, as is the similarity between the orbital energy sequences inferred from P.E. and optical absorption spectra. As a matter of fact, the difference of 2.34 eV between the first two vertical ionizations of VCl_4 at 9.41 eV (d^1) and 11.75 eV $(1\,t_1)$ is not far from the first optical charge-transfer energy of ~ 3.1 eV (69). In other cases, deviations from Koopmans' behavior and corrections for interelectronic repulsion contributions to optical transition energies may lead to more profound and even qualitative discrepancies between the apparent order of energy levels from optical data and the experimental sequence of I.E.s.

The only other well-investigated type of transition metal halides are d^{10} species, including linear AgX and MX_2 (M = Zn, Cd, Hg) (Fig. 9). d-Orbital energies are ex-

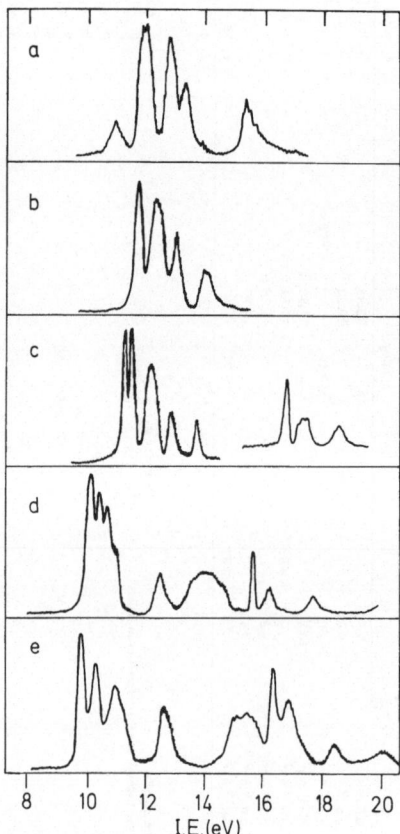

Fig. 9. He(I) P.E. spectra of d^{10} metal halides and related metal-organic derivatives: (a) AgCl (74); (b) $ZnCl_2$ (76); (c) $HgCl_2$ (76); (d) $Hg(CH_3)Br$ (78); (e) $Hg(CF_3)I$ (80)

pected to lie rather deep in the region $\sim 12-20\,\mathrm{eV}$, hence deeper than π- or σ-halogen orbitals. As a consequence, in the case of strong covalent interactions, d-orbitals become the main constituents of bonding rather than antibonding orbitals, and the ligand-field splitting patterns, which are observed in absorption spectroscopy for antibonding d-based orbitals, may appear here in reverse order. This situation is again different from that of d^{10} complexes of low-valent metals as discussed in the previous sections and is again directly observable only in P.E. spectroscopy since these d^{10} systems escape absorption-spectroscopic observation because of both the complete occupancy and the very low energy of their d-shell orbitals.

Silver(I) halides were reported to exhibit two halide ionization bands in the region of lower I.E. (e.g., at 11.03, and 11.90 eV in AgCl, with intensity ratio 1:2) followed by a widely separated triplet ($\sim 3\,\mathrm{eV}$) of d-ionizations, in approximate intensity ratio $2:2:1$, in order of increasing I.E. (74). This is consistent with a picture of strong covalent interaction between halogen p and silver 4 d-orbitals, the halogen orbitals becoming mainly involved in antibonding M.O.s with orbital energies $\sigma_X^* > \pi_X^*$, and the d-orbital splitting being $\delta > \pi > \sigma$, i.e. the anti-ligand field sequence, with very large ligand-field splitting (74).

More extensively investigated are the series of zinc(II), cadmium(II), and mercury(II) halides; the lowest ionizations involve again halide π and σ orbitals, but in different order from silver(I) halides, i.e., with orbital energies $\pi_X > \sigma_X$ such as $\pi_{g,Cl}$ at 11.87, $\pi_{u,Cl}$ at 12.39, $\sigma_{u,Zn\text{-}Cl}$ at 13.07, and $\sigma_{g,Zn\text{-}Cl}$ at 14.10 eV in $ZnCl_2$ (75). Such P.E. spectral patterns are repeated regularly throughout the series, the bromides and iodides being ionized at slightly lower I.E.s, with superimposed effects of spin-orbit splitting ($13, 75-77$). Bond formation has been interpreted assuming that the main covalent interactions occur here between halide p_σ and metal $(n + 1)\,s$, $(n + 1)p$ rather than nd, and the whole situation for linear $X-M-X\,d^{10}$ molecules is strongly reminiscent (even in I.E. numerical values) of that in diatomic $X-X$, except for the obvious presence of one more σ-bonding orbital in the former case. d-Orbital ionization patterns appear at very low orbital energies ($\sim 18.3-19.5$ eV for Zn and $\sim 15-18.6$ eV for Hg compounds) and show a smaller overall d-splitting, which is dominated by spin-orbit rather than ligand-field effects. Thus, in zinc(II) halides ($\zeta_{3d,Zn} \sim 0.15$ eV), the observed d ionizations are assigned, in order, as 19.0 (singlet) $d\sigma$, 19.2 (doublet) $d\,\pi_{3/2} + d\,\delta_{5/2}$, and 19.5 (doublet) $d\,\pi_{1/2} + d\,\delta_{3/2}$ (for $ZnCl_2$); this implies an orbital d-sequence $\sigma > \pi \gtrsim \delta$, with a small total ligand-field splitting $\sim 0.2-0.4$ eV and an ordering which is not compatible with any simple covalent interaction scheme; it has been interpreted "as if" d-orbitals were subject to electrostatic crystal-field effects ($77, 79$). At any rate, interactions involving the $3d$ orbitals of Zn are very weak, in marked contrast with Ag(I) halides ($13, 75, 76$). d-Ionization data are less clear for Cd(II) halides, whereas all linear Hg(II) compounds show characteristic patterns with narrow bands (up to five in number, corresponding to the double group representations $^2\Sigma_{1/2}$, $^2\Pi_{3/2,1/2}$, and $^2\Delta_{5/2,3/2}$) distributed over a range of ~ 2 eV ($\zeta_{5d,Hg} \sim 0.74$ eV). Whenever resolution is not complete, only spin-orbit splitting effects are apparent (78), but the most complete analysis is that by Burroughs et al. (79), who identified all five components on the ground of the sharpness of the ionization peaks due to the nonbonding Δ orbitals and of similarities in vibrational fine structure. For example, $(CH_3)HgCN$ has $^2\Delta_{5/2}$ at ~ 16.2, $^2\Sigma_{1/2}$ at 16.5, $^2\Pi_{3/2}$ at 17.2, $^2\Delta_{3/2}$ at 18.2, and $^2\Pi_{1/2}$ at 18.9 eV (79); $(CH_3)_2Hg$ is less well resolved in this region, having peaks at 14.95 $(^2\Delta_{5/2})$, 15.44 $(^2\Pi_{3/2} + {}^2\Sigma_{1/2})$ and 16.90 eV $(^2\Delta_{3/2} + {}^2\Pi_{1/2})$ (79). The most likely orbital energy ordering compatible with this assignment is $\delta > \pi > \sigma$, which represents a covalency-determined, anti-ligand field sequence different from that of $Zn(II)$ halides, but sharing the basic feature of small overall d-orbital splitting and of weak interactions affecting metal-d orbitals. The good resolution of the d ionization multiplet in most $Hg(II)$ compounds allows further interpretive possibilities, so the energy shift of the strictly nonbonding, easily identified $^2\Delta_{5/2}$ peak is related essentially to the positive atomic charge localized on mercury, and the degree of bond ionicity was estimated by interpolation to be between the values of nonbonding $d_{5/2}$ ionizations of 14.84 eV in atomic mercury and 24.06 in gaseous Hg^+, suggesting figures like 21 % ionic character in $HgCl_2$ and 12% in HgI_2 (78).

The characteristic d-ionization patterns discussed above occur, with but small quantitative shifts, not only in $Hg(II)$ halides, but in all investigated linear bicoordinated $Hg(II)$ species, obviously differing in the region of low I.E. according to the nature of the ligands. There is a vast P.E. literature in this field; by way of example we mention $(CH_3)HgBr$ having a doublet at 10.16–10.43 eV (E, π_{Br}), single peaks at 10.66 $(A_1$, mainly $\sigma_{Hg-C})$ and 12.52 $(A_1$, mainly $\sigma_{Hg-Br})$, and a larger band at 13.9 (E, σ_{CH}) before the d-multiplet at 15.68, 16.11, and 17.64 eV (78). $Wittel\ et\ al.$ investigated several trifluoromethylmercury derivatives (80), e.g., $(CF_3)HgI$ has $E(\pi_I)$ at 9.89–10.42 eV, σ_{Hg-I} at 11.1, σ_{Hg-C} at 12.8, two σ_{C-F} at 15.2 before the d-multiplet at 16.3, 16.9, and 18.1 eV, and the third σ_{C-F} at 19.9 eV (80). A further interesting aspect of the P.E. spectra of $Hg(II)$ compounds is the strong hyperconjugation effect which arises in $Hg(II)$ complexes with ligands containing π systems; interaction with the σ_{Hg-C} bond orbital has thus been observed to decrease strongly the π ionization energy of the ethylene moiety in allylmercuric chloride (81), and of one of the components of the $e_{1g}\pi$ benzene orbitals in benzylmercuric chloride (82).

Halides of posttransition low-valence elements such as $In(I)$, $Tl(I)$, $Sn(II)$, and $Pb(II)$ have also been investigated in some detail by P.E. spectroscopy. d-Orbital energies are usually too low and escape detection by He(I) photoionization [$5d$ is around 19–21 eV in $Tl(I)$ compounds, and $4d$ is around 24–25 eV in $Sn(II)$ compounds], but an interesting additional feature is the photoionization of the nonbonding electron pair mainly formed by metal $(n + 1)$ s. In $Tl(I)$ halides, the P.E. spectrum is very simple, showing only three evident bands in the He(I) region, assigned, for example, in $TlCl$ to $6s(Tl)$ at 9.89, π_{Cl} at 10.38, and σ_{Tl-Cl} at 13.89 eV (83), and likewise in TlF, $TlBr$, and TlI $(83-87)$. The energy position of the $6s$ ionization in thallous halides has been used, like that of $5d$ in mercuric compounds, as a probe for bond ionicity; it turns in fact to be intermediate between the energy of $6s$ in atomic Tl (~ 8 eV) and in gaseous Tl^+ (20.4 eV). The P.E. spectrum of $Tl(I)$ compounds changes drastically under He(II) irradiation: The $5d$ ionization, which is of very small intensity and practically not observed at the limit of the He(I) region, is greatly enhanced and becomes the dominant feature in He(II) spectra, while ns atomic cross sections, including that of thallium $6s$, are sometimes smaller under He(II); halide π ionizations remain still detectable, but with smaller intensity (85).

Sn(II) and Pb(II) halides show multiplets or broad bands of halogen π ionization in the usual energy region ($\sim$ 10–12 eV), as corresponding to the presence of four nondegenerate π_X orbitals in their C_{2v} molecular structure (87); the lowest ionizations are now those of the metal-halogen σ bonds, e.g. at 9.85 eV in $SnBr_2$ and 9.81 eV in $PbBr_2$, reflecting the moderate electronegativity of the metals. Since we are now in a later posttransition group, $(n + 1) s$ energies are lower than in In(I) or Tl(I), and the corresponding ionizations are observed as weak bands at 15.9 eV in $SnCl_2$, 15.6 in $SnBr_2$, and 16.3 in $PbBr_2$ (87).

Nitrogen-, Oxygen-, and Sulfur-Containing Complexes. UPS studies in these fields have been unevenly concentrated on single classes of compounds which have proved either well volatile, or apt to easy assignment, or else particularly amenable to investigation by this technique. A survey of P.E. spectroscopic results covers therefore only partial aspects, does not lend itself to generalization of possible conclusions, and is best performed by analyzing the P.E. spectroscopic behavior class by class of investigated compounds since the observed P.E. spectra are usually dominated by the ionization patterns of the various types of ligands, possibly modified on complexation.

Amide and Related Complexes (Table 5). Dialkylamido ligands bind to tetravalent elements giving pseudotetrahedral $M(NR_2)_4$ ($R = CH_3$, $M = C, Si, Ge, Sn, Ti, Zr$, Hf, V; $R = C_2H_5$, $M = Ti, Zr, Hf$) and octahedral $W[N(CH_3)_2]_6$, whose P.E. spectra were investigated by *Gibbins et al.* (24). The lowest ionizations are (except for the V compound) those of the nitrogen lone pairs; these occur at 8.85 eV in the free ligand $(CH_3)_2NH$ and at 8.68 in $(C_2H_5)_2NH$ (+ I effect of the alkyl substituents) and appear in the complexes shifted to lower energies (between $\sim$ 7 and $\gtrsim$ 8 eV), mainly as a consequence of the negative character of ligand R_2N^-, and split into partially resolved patterns (effective coordination symmetry D_{2d}). Metal-nitrogen σ ionizations start in the region 10.3–11.2 eV (tetrahedral t_2 component) for the dimethylamido and 9.5–10.0 eV for diethylamido species {except the compound $C[N(CH_3)_2]_4$ at $\sim$ 12.8 eV}, and $\sigma_{CH} + \sigma_{CN}$ beyond 12.2 eV (24). The V compound is the only one in this series having d electrons; its d^1 ionization is clearly evident in the extra band at $\sim$ 6.2 eV (24), i.e., d orbitals are ionized considerably before ligand orbitals. A similar situation seems to occur in the tetraalkyl species $M[CH_2X(CH_3)_3]_4$ ($X = C$, $M = $ Ti, Zr, Hf, Ge, Sn, Cr; $X = Si$, $M = $ Ti, Zr, Hf, Sn, Pb, Cr), where the σ_{M-C} ionizations (tetrahedral component t_2) occur at fairly constant I.E. at 8.2–9.0 eV, followed by σ_{C-C} (or σ_{Si-C}) ionizations between 10.3 and 11.3 eV and by a large and broad ionization band of CH bonds starting around 12.2–13.6 eV (88, 89). The only non-d^0 systems in these series are the chromium complexes, which exhibit an extra band at 7.25 eV, again assigned to the partly filled shell, which is here a d^2 configuration (89). Other d^0 systems investigated by UPS include octahedral $W(CH_3)_6$ and trigonal-bipyramidal $Ta(CH_3)_5$, for which conflicting assignments were proposed (90, 91).

Table 5

Compounds	Formal metal electronic configuration	d	σ_{M-C}		σ_{C-C} (or σ_{Si-C})	σ_{C-H}				References[b]
ClCH$_2$SiMe$_3$					10.17	14.3	11.0[c]	12.24[f]		89
HCH$_2$CMe$_3$					11.3	12.5				88
HCH$_2$SiMe$_3$					10.6	13.1	14.0			88
Ti(CH$_2$CMe$_3$)$_4$	d^0		8.33		11.35	12.59				88
Zr(CH$_2$CMe$_3$)$_4$	d^0		8.33		11.28	12.50				88
Hf(CH$_2$CMe$_3$)$_4$	d^0		8.51		11.40	12.54				88
Ge(CH$_2$CMe$_3$)$_4$	d^0		9.01		10.28	12.25				88
Sn(CH$_2$CMe$_3$)$_4$	d^0		8.58		11.16	12.37				88
Ti(CH$_2$SiMe$_3$)$_4$	d^0		8.58		10.46	13.35				88
Zr(CH$_2$SiMe$_3$)$_4$	d^0		8.64		10.28	13.22				88
Hf(CH$_2$SiMe$_3$)$_4$	d^0		8.58		10.27	13.36				88
Sn(CH$_2$SiMe$_3$)$_4$	d^0		8.71		10.3	13.2				89
Pb(CH$_2$SiMe$_3$)$_4$	d^0		8.14	8.86	10.3	13.2				89
Cr(CH$_2$CMe$_3$)$_4$	d^2	7.25	8.37		11.0	12.2	13.2	14.5	17.0	89
Cr(CH$_2$SiMe$_3$)$_4$	d^2	7.26	8.69		10.4	13.6				89

Compounds	Formal metal electronic configuration	S_M^2	σ_{M-C}	References[b]
Hg[CH(SiMe$_3$)$_2$]$_2$	$5d^{10}$		8.12	92
Ge[CH(SiMe$_3$)$_2$]$_2$	$3d^{10}4s^2$	7.75[g]	8.87[h]	92
Sn[CH(SiMe$_3$)$_2$]$_2$	$4d^{10}5s^2$	7.42[g]	8.33[h]	92
Pb[CH(SiMe$_3$)$_2$]$_2$	$5d^{10}6s^2$	7.25[g]	7.98[h]	92

Compounds	Formal metal electronic configuration	N lone pairs					$a_1(s^2 M)$	σ_{M-N}	References[b]
(SiMe$_3$)$_2$NH		8.79[i]		[10.23	10.54	10.74[j]	[11.26	11.59[k] ~15[l]	92, 93, 25
(SiMe$_3$)(CMe$_3$)NH		8.41[i]							92, 93, 25
Zn[N(SiMe$_3$)$_2$]$_2$	$3d^{10}$		8.50					9.55	92, 93, 25
Hg[N(SiMe$_3$)$_2$]$_2$	$5d^{10}$		8.33					9.38	92, 93, 25
Ge[N(SiMe$_3$)$_2$]$_2$	$3d^{10}4s^2$	7.71[h]	8.99[g]				8.68		92, 93, 25
Sn[N(SiMe$_3$)$_2$]$_2$	$4d^{10}5s^2$	7.75[h]	8.85[g]				8.38	9.50	92, 93, 25
Pb[N(SiMe$_3$)$_2$]$_2$	$5d^{10}6s^2$	7.92[h]	8.81[g]				8.16	9.39	92, 93, 25
Ge[N(SiMe$_3$)(CMe$_3$)]$_2$	$3d^{10}4s^2$	7.24[h]	8.61[g]				8.27		92, 93, 25
Sn[N(SiMe$_3$)(CMe$_3$)]$_2$	$4d^{10}5s^2$	7.26[h]	8.47[g]				7.90	9.33	92, 93, 25
Pb[N(SiMe)(CMe$_3$)]$_2$	$5d^{10}6s^2$	7.26[h]	8.49[g]				7.69	9.00	92, 93, 25

Compounds	Formal metal electronic configuration	N lone pairs		σ_{M-N}	Ligand orbitals			σ_{C-H}	References[b]
Sc[N(SiMe$_3$)$_2$]$_3$	d^0	8.18[m]	8.62[n]	9.45	10.09		10.98	~14	25
Ti[N(SiMe$_3$)$_2$]$_3$	d^1	8.14[m]	8.70[n]		10.22		11.07	~13.5	25
Cr[N(SiMe$_3$)$_2$]$_3$	d^3	8.07[m]	8.76[n]		10.16	11.04	11.21	~13.5	25
Fe[N(SiMe$_3$)$_2$]$_3$	d^5	7.88[m]	8.74[n]		10.13	11.06		~13.5	25
Ga[N(SiMe$_3$)$_2$]$_3$	d^{10}		8.39		10.09		11.07	~13.8	25
In[N(SiMe$_3$)$_2$]$_3$	d^{10}		8.30		10.22		11.11	~13.8	25

Compounds	Formal metal electronic configuration	d	N lone pairs				$t_2(\sigma_{M-N})$		$\sigma_{C-H} + \sigma_{C-N}$			References[b]
Me$_2$NH			8.85[i]									24
C(NMe$_2$)$_4$			7.19	8.43	9.20				~13.0			24
Si(NMe$_2$)$_4$				8.69			11.21		~12.8			24
Ge(NMe$_2$)$_4$	d^0			8.48			11.16		~13.5			24
Sn(NMe$_2$)$_4$	d^0			7.67			10.84		~12.5	13.5		24
V(NMe$_2$)$_4$	d^1	6.2	7.08	7.60	8.28		10.41		12.3	13.8	14.5	24
Ti(NMe$_2$)$_4$	d^0		7.13	7.36	7.75	8.00	10.32		12.8	13.2	14.5	24
Zr(NMe$_2$)$_4$	d^0		7.23	7.54	7.92	8.14	10.44		12.6	13.5	15.0	24
Hf(NMc$_2$)$_4$	d^0		7.50	7.82	8.05	8.34	10.56		12.5	13.8	14.8	24
Et$_2$NH	d^0		8.68[i]									24
Ti(NEt$_2$)$_4$	d^0		6.83	7.10	7.47	7.75	9.78		12.3			24
Zr(NEt$_2$)$_4$	d^0		6.76	6.98	7.35	7.54	9.55		12.2			24
Hf(NEt$_2$)$_4$	d^0		7.15	7.35	7.68	7.91	9.97		12.0			24
W(NMe$_2$)$_6$	d^0		6.73	7.92			9.55	9.95	11.4	12.5	14.0	24

Compounds	Formal metal electronic configuration	σ_{M-C}				σ_{C-H}			References[b]
TaMe$_5$	d^0	~8.6[o]	9.2[p]	10.2[n]	11.3[o]	11.2	11.8	12.9	90
WMe$_6$	d^0	~8.9[q]	9.2[r]	10.0[s]		11	12	13	90, 91

152

Compounds	Formal metal electronic configuration	Ionization energies (eV)[a]										References[b]
		d			Ligands π-orbitals			Ligands σ-orbitals				
Ni(C₃H₅)₂	d^8	7.85[t]	8.17[t]	8.59[u]	9.48[v]	10.44[w]	11.56[u]	12.75	14.20	15.73	17.67	10, *117*
Pd(C₃H₅)₂	d^8	7.59[t]	8.74[u]	9.22[x]	9.52	9.75[v]	10.43[w] 11.56[u]	12.83	14.13	15.61	(17.66)	116
		Ligands orbitals										
H₂TPP		6.39	6.72		7.71	8.86	11.63	13.57		14.32		94
MgTPP	d^0	6.48			7.84	8.91	11.33	13.52		14.08		94
MnTPP	d^5	6.44	6.61		7.66	8.80	11.42	13.45		14.09		94
FeTPP	d^6	6.50	6.80		8.02	8.92	11.65	13.72		14.31		94
NiTPP	d^8	6.44	6.62		8.04	8.93	11.72	13.77		14.54		94
CuTPP	d^9	6.49	6.66		7.77	8.78	11.37	13.37		14.03		94
ZnTPP	d^{10}	6.42	6.62		7.76	8.73	11.52	13.49		14.25		94

[a] Shoulders in parentheses.	[k] σ Si–N.	[s] e_g-type sublevel.
[b] Numerical data are taken from the references in italics.	[l] σ C–H.	[t] a_g-type sublevel + b_g-type sublevel.
[c] Chlorine π-orbital.	[m] a_2-type sublevel.	[u] a_g-type sublevelL
[f] σ C–Cl.	[n] e-type sublevel.	[v] a_u-type sublevel.
[g] a_1-type sublevel.	[o] a_1'-type sublevel.	[w] b_u-type sublevel.
[h] b_2-type sublevel.	[p] a_2'-type sublevel.	[x] b_g-type sublevel.
[i] Unsplit N lone pair.	[q] t_{1u}-type sublevel.	
[j] σ Si–C.	[r] a_{1g}-type sublevel.	

Completely different is the P.E. spectroscopic behavior of disilylamido complexes $M\{N[Si(CH_3)_3]_2\}_3$ (M = Sc, Ti, Cr, Fe, Ga, In) (Fig. 10) where the lowest ionizations are in all cases those of the nitrogen lone pairs [8.79 in the free disilylamine ligand (*25, 92, 93*) and ~7.9–8.7 eV in the complexes], and the d^n configurations of the Ti, Cr, and Fe compounds do not give rise to detectable separate ionizations before or amid the ligand ionizations. The different coordination symmetry, close to D_{3h}, is no sufficient motive for such drastic change in P.E. behavior from e.g. dialkylamidocomplexes, and the explanation suggested by *Lappert et al.* (*25*) is that N → Si π-bonding lowers the negative charge on nitrogen (i.e., 7.9–8.7 eV in the complexes) so as to counterbalance M → N π-backdonation, which remains more pronounced in dialkylamidocomplexes (I.E. of N ~ 7.0–8.0 eV). As a consequence, the atomic charges on the central metals in dialkylamido complexes are lower than in disilylamido complexes; d-orbital ionizations occur at lower I.E. values and are therefore seen as separate bands only in the former case (*25*). This is, however, only a partial and simplified explanation. Actually, the position and distinguishability of d^n ionizations depend on several more factors, as we shall discuss in the next section. However, for early transition metals, and in comparable coordination environments, the mechanism proposed by *Lappert et al* (*25*) is probably the discriminating one, although the argument can be reformulated by considering that not only the higher positive charge on the central metal, but also the mixing of metal d-orbitals with the N–Si π-system makes the d-orbital energies too low to be observed as distinct ionization bands before ligand ionizations in the disilylamido complexes. Similar conclusions are suggested by the P.E. spectra of metal mesotetraphenylporphine complexes (*94*), where the planar quadratic N_4 ligand system exhibits a well-defined series of narrow ionization peaks at ~ 6.4, 6.7, 7.7, 8.9, 11.6, 13.6, and 14.3 eV plus or minus few hundredths of an electron volt, irrespective of the central metal ion, with no extra ionization bands attrib-

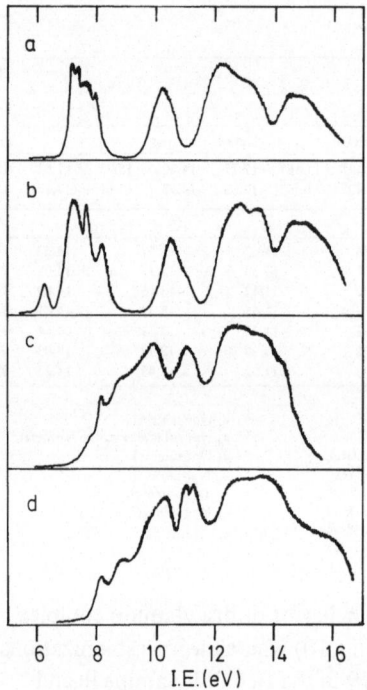

Fig. 10. He(I) P.E. spectra of alkyl- and silyl-amido metal complexes: (a) Ti(NMe$_2$)$_4$ (24); (b) V(NMe$_2$)$_4$ (24); (c) Sc[N(SiMe$_3$)$_2$]$_3$ (25); (d) Cr[N(SiMe$_3$)$_2$]$_3$

utable to d-ionizations in the complexes of bivalent Mn, Fe, Ni, Cu, and Zn. There are two obvious reasons for the lack of distinct d-ionization bands, the presence of a system of highly negative ligand nitrogen atoms at unusually high orbital energies (up to approx. -6.4 eV) which are likely to be ionized before any d^n configuration, and the possibility of extensive conjugation between metal d-orbitals and the ligand π-system, introducing a substantial ligand participation to the M.O.s mainly based on metal d atomic functions.

Evidence for substantial $N \rightarrow Si\ p_\pi - d_\pi$ interaction was claimed by Harris et al. (92) also in the P.E. spectra of the amido complexes M{N[Si(CH$_3$)$_3$]$_2$}$_2$ (M = Zn, Hg, Ge, Sn, Pb) and M{N[Si(CH$_3$)$_3$] [C(CH$_3$)$_3$]}$_2$ (M = Ge, Sn, Pb). The same authors investigated also the P.E. spectra of the bicoordinated alkyl derivatives M{CH[Si(CH$_3$)$_3$]$_2$}$_2$ (M = Hg, Ge, Sn, Pb), where it is interesting to note that no bands were observed for the ionization of the d-electrons of mercury (92), contrary to the findings in most other mercury complexes (see Table 4.).

Acetylacetonates and Related Complexes (Table 6). Metal β-diketonates and related functional derivatives have been extensively investigated in UV-P.E. spectroscopy. Their spectra are similar to those of the free β-diketone ligands, whose spectral patterns can always more or less directly be traced back in the spectra of the metal complexes, of which they often represent the major and more characteristic part. d-

Ionizations are evident before ligand ionizations with early transition metals Ti(**III**), V(**III**), or Cr(**III**), not evident at all or masked under ligand ionization bands with late transition metals such as Cu(**II**) or Zn(**II**), and highly intermixed with the first ligand ionizations for intermediate transition metal configurations such as d^6 of Co(**III**) or d^8 of Ni(**II**). (Fig. 11). This type of behavior is probably of general occurrence and is in fact found also in other classes of coordination compounds, especially with sulfur ligands.

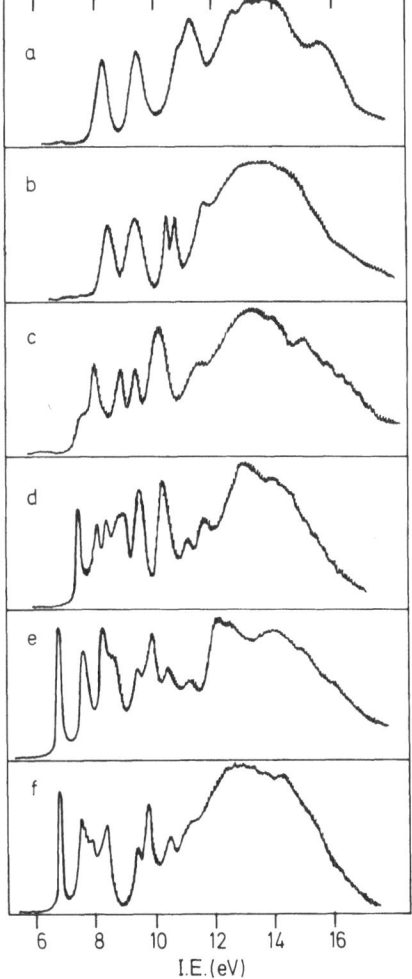

Fig. 11. He(I) P.E. spectra of metal β-diketonates and related species: (a) Be(acac)$_2$ (*95*); (b) Zn(acac)$_2$ (*96*); (c) Cr(acac)$_3$ (*95*); (d) Ni(acac)$_2$ (*96, 98*); (e) Ni(acacS$_2$)$_2$ (*98*); (f) Ni(acac-en) (*99*)

Table 6

Compounds	Formal metal electronic configuration	d	π_3	n_-	n_+	$\pi_2 + \sigma_{C-C} + \sigma_{C-H}$	References[b]
Hacac			9.18	9.74	12.68	> 13.3	95
Htfac			9.92	10.53	13.27	> 15.2	95
Hhfac			10.74	11.25	14.03	> 15.0	95
Be(acac)₂	d^0		8.41	9.67	10.48		96
Be(hfac)₂	d^0		10.39	11.66	12.96	> 13.0	95
Mg(hfac)₂	d^0		10.28	11.18	12.31	> 13.4	96
Al(tfac)₃	d^0		9.22	10.35	11.49		95
Al(hfac)₃	d^0		10.33	11.47	12.64	> 14.7	95
Ga(hfac)₃	d^0		10.19	10.82, 11.13	12.37	> 13.6	95
Sc(hfac)₃	d^0		10.13	10.87, 11.28	12.05	> 13.0	95
Ti(hfac)₃	d^1	7.94	10.24	10.96, 11.39	12.24	> 13.5	95
V(hfac)₃	d^2	8.68	10.10	10.96	12.20	> 13.5	95
Cr(acac)₃	d^3	7.46	8.06	8.96, 9.48	10.26		95
Cr(tfac)₃	d^3	8.58	9.12	10.01, 10.54	11.40	> 13.6	95
Cr(hfac)₃	d^3	9.57	10.18	11.10, 11.61	12.44	> 13.5	95
Mn(acac)₃	d^4	(7.32)	8.14	10.94, 11.41	12.31		95
Mn(hfac)₃	d^4	(9.2)	10.03	(8.93), 9.22	11.26	> 13.5	95
Fe(acac)₃	d^5	8.10	10.94	(9.92), 10.40	12.25 / 10.16	> 13.3	95
Fe(tfac)₃	d^5	9.18		11.06, 11.65	12.50 / 11.4	> 13.2	95
Fe(hfac)₃	d^5	10.13		11.30	11.19		95
Ru(hfac)₃	d^5	8.85, 9.07	10.30	11.06, 11.65		> 13.5	96, 98, 97
Mn(CO)₄(hfac)	d^6	8.11	9.31	10.31, 11.15	12.56		98
Co(acac)₃	d^6	7.52, 8.03	8.49	8.99, 9.54	10.34	> 13.5	96
Co(hfac)₃	d^6	9.73, 10.13	10.73	11.15, 11.75	12.56	> 13.9	96
Ni(acac)₂	d^8	7.41, 7.89, 8.15	8.38	8.75, 9.26	10.06		96, 98, 97
Ni(tfac)₂	d^8	8.25, (8.75), 8.92	9.30	9.65	10.98	> 13.5	96, 97
Ni(hfac)₂	d^8	9.35, 9.84		10.67, 11.11	12.01		96
Cu(acac)₂	d^9		8.33	8.69, 9.29	10.38[c]		96, 97
Cu(hfac)₂	d^9		10.18	10.60, 11.11	12.20[c]		96
Zn(acac)₂	d^{10}		8.46	9.22, 10.39	9.84, 10.72		96
Zn(hfac)₂	d^{10}		10.25	11.17, 12.53	10.88, 12.78		96
Ni(Sacac)₂	d^8	6.99, (7.54), 7.63	8.44	8.84	9.46		98
Ni(Stfac)₂	d^8	7.80, 8.51	9.28	9.58	10.33, 9.68		98
Ni(SacSac)₂	d^8	6.92, (7.63), 7.73	8.31	8.58	9.26, 10.06		98
Ni(S₂tfac)₂	d^8	7.65, (8.38), 8.58	9.08	9.38	9.98, 10.36, 10.68		98
H₂BAE		7.71	(7.90)	8.40	8.78	> 11.7	99
NiBAE	d^8	7.60, 7.90[f]	6.80[g]	8.40	8.78	10.47, 10.60	99
PdBAE	d^8	6.88[g], 7.54, 8.25	8.46	9.18[i]	9.30[h]	9.68	99
CuBAE	d^9	7.00	7.57	8.65[j], 8.81	9.68, (10.1)[k]	> 11	99

a Ionization energies (eV)

		d			Ligand + σ(Mo–O) orbitals						
$Mo_2(HCO_2)_4$	$[d^4]_2$	7.5[v]	9.5[w]		11.0[x]	11.6	12.7	14.9	16.1	16.7	119
$Cr_2(H_3CCO_2)_4$	$[d^4]_2$	8.65[y]	(9.1)[y]	(10.2)[b]	10.51[z]	11.04	12.08	13.33	14.11	15.66	120
$Mo_2(H_3CCO_2)_4$	$[d^4]_2$	6.8[v]	8.7[w]		10.4[x]	11.1	11.95	13.3	14.1	15.3	119
$Mo_2(Me_3CCO_2)_4$	$[d^4]_2$	6.7[v]	8.5[w]		10.1[x]	10.8	13.2	15.8	16.9		119

		d				π_s					n_O	
Hdtp							9.1[l]			10.2[m]	~10.8	*101*, 102, 103
$Crdtp_3$	d^3		7.56				8.44		9.2	10.15	11.19	102, *103*
$Fedtp_3$	d^5						8.65		9.3	10.3		102, *103*
$Codtp_3$	d^6		8.15			8.7		9.05	9.7	10.5	11.3	102, *103*
$Nidtp_2$	d^8	7.24		8.37			9.2		9.9	10.3	~10.8	*101*, 102, 103
$Pddtp_2$	d^8	7.90			8.70[n]				9.7	10.4	~10.8	101
$Ptdtp_2$	a^1	7.60		8.80			9.2		10.0	10.4	~10.8	101
$Zndtp_2$	d^{10}						8.87		9.94	10.65	11.41	102, *103*

		d			Ligands π			σ M–S			σ P–S		π_F (+ σ_F)		σ_{PF} (+ π_F)	
HS_2PF_2						10.47[o]		11.52[p]	13.05[q]		[13.50	14.80][r]	16.4	17.9	19.1	105
$Cr(S_2PF_2)_3$	d^3		9.41			10.02	10.93	11.98	13.04		13.34	15.10	16.4	17.8	19.1	105
$Mn(S_2PF_2)_2$	d^5		9.38			10.08	10.6	11.30	11.50	12.44	13.00	15.14	16.2	17.55	19.0	105
$Co(S_2PF_2)_2$	d^7	(9.18)	9.62	9.98				(12.3)	(13.2)		(14.0)	15.02	(16.0)	(17.7)		105
$Ni(S_2PF_2)_2$	d^8	(8.62)	8.76	(8.94)	9.84	10.45	10.58	11.44	12.84	(13.04)	13.47	14.76	16.17	17.85	18.95	105
$Zn(S_2PF_2)_2$	d^{10}					10.12	10.34	11.70		13.14	13.74	15.15	16.10	17.45	19.35	105

		d				π_3		σ M–S (n_-)		π_2		σ M–S (n_+)				
$Crdtc_3$	d^3		7.02				7.48		8.13		8.42			8.83		108
$Fedtc_3$	d^5				7.75				8.19			8.39				108
$Fe[S_2CN(CH_3)_2]_3$	d^5				7.72				8.04			8.37				109
$Codtc_3$	d^6		6.67				7.35		8.25					9.45		108
$Nidtc_2$	d^8	6.95		7.13			7.84	8.10[s]		8.66[t]				9.07		108
$Cudtc_2$	d^9				7.63	(7.93)				8.13						108
$Zndtc_2$	d^{10}					(7.69)	(7.85)	(8.13)			8.31		8.75		9.41	108
$Fe[S_2CN(CH_3)_2]_2$[u]	d^6				(7.79)				(8.14)			8.51	(8.90)			109
$Fe[S_2CN(C_6H_2)_2]_2$[u]	d^6				7.58				(7.95)			(8.26)	9.12			109

[a] Shoulders in parentheses.
[b] Numerical data are taken from the references in italics.
[c] Metal "d" + n_+ orbitals.
[f] Metal "d" + π_3 orbitals.
[g] π_3 orbitals.
[h] Metal "*d*" + n_- orbitals.
[i] Metal "*d*" + π_3 + n_- orbitals.
[j] n_- orbitals.
[k] Metal "*d*" + n_- + n_+ orbitals.
[l] n_s (P=S).
[m] n_s (PSH).
[n] Metal "*d*" + π_s orbitals.
[o] $\pi_{P=S}$.
[p] σ_{PSH}.
[r] $\sigma_{PS} + \sigma_{S-H}$.
[s] $b_{3u}(n_-)$.
[t] $b_{1g}(n_-) + \pi_2$.
[u] Other I.E. values relative to analogous compounds with different substituent groups on the nitrogen are reported in Ref. 109.
[v] $^2B_{2g}$ (δ[Mo–Mo])
[w] ^2E_u (π[Mo–Mo])
[x] mixed with σ[Mo–Mo]
[y] $9b_{1u} + 3a_u$ (δ* + σ*[Cr–Cr])
[z] σ + δ[Cr–Cr]

The P.E. spectra of β-diketones, which occur in the gas phase in enol form, hence directly comparable with the metal chelate structures, consist of a sequence of two or three bands, e.g., at 10.74, 11.25, and 14.03 eV in hexafluoroacetylacetone (95). This accounts, in order of increasing I.E., for the highest occupied π orbital of the chelate ring (π_3, with positive coefficients on both oxygen atoms) and for the negative and the positive linear combinations of the in-plane lone pairs of oxygen n_- and n_+, of which the latter is σ-bonding to H in the enol ring (95). In the same energy region also the ligand π_2 ionization is expected (95). The fluorosubstituted acetylacetones tfacH = $CF_3COCH_2COCH_3$ and hfacH = $CF_3COCH_2COCF_3$ have P.E. spectra similar to acetylacetone itself, up to a parallel shift of all observed I.E.s, as a consequence of the strong $-$ I effect of fluorine (the so-called "fluorine effect", which is observed in many cases of P.E. spectra of organic substances).

β-diketonates of nontransition, or of d^0 transition metals exhibit in their P.E. spectra the same sequence of bands, empirically termed A, B, C, D, slightly shifted to lower I.E.s because of the negative charge of β-diketonate anions, which is only in part redistributed to the central metal atoms in the complexes, and the lowering in orbital energy as the β-diketonate orbitals acquire bonding character in the coordination sphere is not so large as to reverse the overall energy shift from the free ligand (enol form). Moreover, splitting due to interligand interactions becomes sometimes evident, although not generally resolved; thus in pseudooctahedral tris-β-diketonates (D_{3d}) the set of three π_3 orbitals, usually associated with band A, transforms like $e + a_2$, the n_- orbitals (band B) like $e + a_2$, and n_+ (band C and perhaps also D) like $e + a_1$, while in pseudotetrahedral bis-β-diketonates (D_{2d}) π_3 and n_- transform like e, while n_+ gives $a_1 + b_2$ (95, 96). As an example, Be(acac)$_2$ has band A at 8.41, B at 9.67, and C at 11.13 eV; Al(hfac)$_3$ has A at 10.33, B at 11.47, C at 12.64, and D at 13.62 eV (95). The same spectral patterns are exhibited by d^{10} metal β-diketonates, e.g., Zn(acac)$_2$ exhibits P.E. peaks assigned as π_3 (8.46 eV), n_- (9.22), n_+ (split into two components, b_2 at 10.39 and a_1 at 10.72 eV (96)), without evidence for separate bands due to d ionizations or even for their possible localization under the envelopes of the other bands; the d^9 planar system Cu(acac)$_2$ has again a spectrum very similar to that of Zn(acac)$_2$, despite the different coordination geometry, and again without evidence for separate bands due to d ionization, where however He(II) spectra suggest, by the observed increase in intensity, that d ionizations are masked under the profiles of the bands at 8.33 and 10.38 eV (97). As to other d^n systems in the β-diketonate series, the d^1 configuration of Ti(acac)$_3$, d^2 of V(acac)$_3$, and d^3 of Cr(acac)$_3$

give rise to extra bands in the region of low I.E., which are easily identified as due to d-ionizations [e.g., at 7.94 eV in Ti(hfac)$_3$, 8.68 eV in V(hfac)$_3$, and 9.57 eV in Cr(hfac)$_3$] (95). On passing to d^4 systems such as Mn(hfac)$_3$, d^5 such as in Fe(acac)$_3$ and Fe(dpm)$_3$, d^6 of Co(acac)$_3$, and d^8 of Ni(acac)$_2$ (which behaves as a monomeric quadratic low-spin species in the gas phase), the spectral patterns in the region of lower I.E. become far more complex and only partially resolved. The obvious implication is that d-orbital ionizations occur at energies comparable with those of π_3 and n_- coordination orbitals, which leaves, however, questions as to the degree of mixing between d- and ligand-based orbitals and as to the actual sequence of orbital energies, particularly in view of the doubts raised on the validity of Koopmans' theorem in open-shell systems and in systems comprising orbitals with different admixtures of d-character (7–12). As a result, no one-to-one assignment is usually presented in the literature for such complexes; only for Ni(acac)$_2$ were proposals of a more detailed assignment attempted (96, 98). The occurrence of narrow bands at 7.41, 7.89, and 8.15 eV (98), at much lower ionization energy than in the free ligand or in any d^0 complex of the same ligand, led both *Brittain* and *Disch* (96) and *Cauletti* and *Furlani* (98) to assign them as d-orbitals ionizations; however, subsequent He(II) measurements showed that the level ionized at 7.41 eV has a significant admixture of ligand character (probably π_3), d-character being more concentrated in the bands with maxima at 7.89 and 8.15 eV (97).

Complexes of several derivatives of β-diketones have also been investigated in P.E. spectroscopy, and their behavior can be compared with the related, or parent, β-diketonates. Thio- and dithiosubstituted acetylacetonates resemble strongly the corresponding oxygen ligands in their behavior as ligands and also in the P.E. spectra of the complexes formed with transition metals. Unfortunately, most work has been done on complexes of Ni(II), whose assignment is still subject to considerable uncertainty, as discussed above. Replacement of oxygen through sulfur results in slight shifts (up to ~ 0.6 eV) of the ionization energies of the lowest bands to smaller values, while the general spectral patterns remain substantially unchanged, apart from a somewhat better resolution in the spectra of the sulfur analogs. Therefore, both the proposed assignment and the doubts that can be raised on the interpretation of the P.E. spectra of Ni(II) thio- and dithioacetylacetonates resemble closely those for the oxygen analogs, although the particularly low value of the first ionization makes their assignment to d-orbitals (or a large d-contribution to the ionized orbitals) more plausible. As an example, the lowest ionizations in Ni(II)-bis-dithioacetylacetonate occur at 6.92, 7.63 sh, and 7.73 eV and are assigned by *Cauletti* and *Furlani* (98) to d-ionizations (7.41, 7.89, and 8.15 eV in Ni(II)-bis-acetylacetonate), followed by a strong band at 8.31 (8.38) assigned as π_3 and by two widely split doublets at 8.58–9.26 (8.75–9.26) and 9.68–10.06 (10.06, broad) assigned to the n_- and n_+ components, respectively.

The Schiff base of acetylacetone with ethylenediamine BAEH$_2$ gives stable quadratic chelates with several transition metals, whose P.E. spectra have been reported by *Condorelli et al* (99). The free ligand has P.E. bands at 7.71, 7.90 sh, and 8.78 eV, attributable in order of increasing I.E., to π_3-like orbitals and to n_- and n_+-like combinations of O and N lone pairs; replacement of O through N therefore considerably lowers the ligand ionization energies. The complexes of Ni(II), Pd(II), and Cu(II) all show the same P.E. spectrum, just shifted to slightly lower I.E. values, e.g., 6.80, 7.60, and 8.40 eV in NiBAE, with no extra bands attributable to d-ionizations, except perhaps within the complex structure of some of the bands at higher I.E., thus yielding simpler and clearer evidence for the fact that d-ionizations do not occur, in the above d^8 and d^9 configurations, before ligand ionizations. This is consistent with the relatively low ionization energy of the valence orbitals already in the free ligand (99). Of the posttransition metals, thallium(I) β-diketonates, investigated together with the dialkylthallium(III) β-diketonates by *Cauletti et al.* (100), show the usual patterns of ligand ionizations and in addition the interesting feature of a narrow band at low I.E. [e.g., 9.30 eV in Tl(acac)] due to the $6s^2$ electron pair of Tl(I), analogous to other Tl(I) compounds, as discussed in a previous section.

Dimeric Cr(II) and Mo(II) carboxylate complexes, investigated by *Green et al.* (119, 120) show low-energy multiple ionization patterns of the $[d^4]_2$ system, assigned to metal–metal σ, π and δ bonding and antibonding orbitals.

Sulfur-Containing Metal Chelates (Fig. 12). The P.E. spectra of chelating sulfur ligands such as dithiophosphates or dithiocarbamates are known either from the free acids, e.g. $HS_2P(OC_2H_5)_2$ (101, 102, 103) or from their esters, e.g. $(CH_3)_2NCS_2CH_3$ (104) and consist, in the region of low I.E., of peaks corresponding to the successive ionizations of sulfur lone pairs, of sulfur π orbitals extended over the conjugated

$X\begin{smallmatrix}\diagup S \\ \diagdown S\end{smallmatrix}$ moiety, and of S–X σ-bonding electrons. The correspondence between the

P.E. spectra of the complexes and of the free ligand acids or esters is not very close, because of the inequivalence of the two sulfur atoms in the latter species, and because of the near-degeneracy of the energies of both lone pairs on the X=S sulfur atoms in the free ligands, which become an in-plane lone pair (potentially σ-bonding) and a π_S component in the metal chelate rings. The simplest structures, and the most thoroughly resolved P.E. spectra, are those of the difluorodithiophosphate metal complexes $M(S_2PF_2)_n$, where the absence of organic substituents and of the corresponding broad bands of σ_{CH} and σ_{CC} ionizations in the region 12–16 eV allow observation of well-resolved P.E. bands in the whole He(I) spectral region (105). HS_2PF_2 has an unresolved doublet of $n_S(P=S)$ ionizations at 10.47 eV followed by $\sigma_{P=S}$ at 11.52, $n_S(PSH)$, and $\sigma(P–S–H)$ bonds between 13.05 and 14.80, and the π and σ fluorine orbitals between ~ 16 and 21 eV, as observed in many P–F compounds, e.g., PF_3 (30, 33) or OPF_3 (106). The latter ionization patterns of fluorine-based orbitals are repeated almost unchanged in the metal complexes of $S_2PF_2^-$. At lower I.E. values a system of six bands is evident in all metal complexes investigated in P.E. spectroscopy and is assigned, in order of increasing I.E., to the two π pairs present in each chelating ring, to the two lone pairs of S which become M–S σ-bonding, and to the two X–S σ-bonding pairs. For example, in $Cr(S_2PF_2)_3$ the six bands occur at 10.02, 10.93, 11.98, 13.04, 13.34, and 15.10 eV; the bands are sometimes broadened, but not ac-

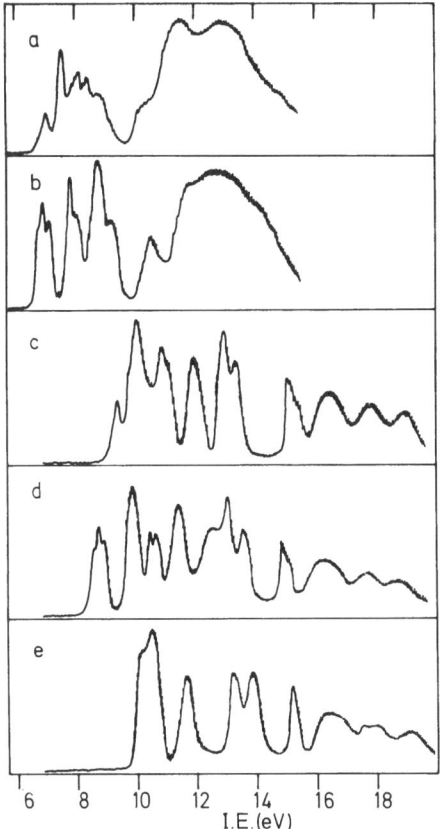

Fig. 12. He(I) P.E. spectra of some sulfur-containing metal chelates: (a) Cr(dtc)$_3$ (*108*); (b) Ni(dtc)$_2$ (*108*); (c) Cr(S$_2$PF$_2$)$_3$ (*105*); (d) Ni(S$_2$PF$_2$)$_2$ (*105*); (e) Zn(S$_2$PF$_2$)$_2$ (*105*)

tually split, because of interligand interactions. Besides such system of ionization bands of ligand-based orbitals, additional bands at lower I.E. are observed in the Cr(**III**), Mn(**II**), Co(**II**), and Ni(**II**) complexes [but not for Zn(**II**)] and assigned as *d*-ionizations, the d^3 configuration of Cr(S$_2$PF$_2$)$_3$ being ionized as a narrow peak at 9.41 eV, the d^5 high-spin configuration of Mn(S$_2$PF$_2$)$_2$ at 9.38 eV as a broad foot at the low energy side of the first ligand band, while in Co(S$_2$PF$_2$)$_2$ the *d* bands merge into a complex broad envelope with the ligand band at 9.62 eV, and in Ni(S$_2$PF$_2$)$_2$ there is a well-separated multiple, partly resolved band with maximum at 8.76 eV. The case of the Ni(**II**) compound is particularly interesting since it represents the best evidence hitherto achieved for occurrence of *d* ionizations before ligand ionizations in complexes of Ni(**II**), in contrast with the confused situation or with easier ligand ionization, e.g., in Ni(**II**)-β-diketonates or Schiff base complexes, although here too some doubt can be cast on the assignment of the first P.E. bands of Ni(S$_2$PF$_2$)$_2$ if one assumes that a profound mixing can take place between metal-*d* and sulfur orbitals. That *d*-ionizations are the first ones to occur in Ni(S$_2$PF$_2$)$_2$ appears plausible in view

161

of the strong inductive effect of fluorine, which lowers the orbital energies of the $S_2PF_2^-$ ligand moieties, so as to leave d-based levels probably at higher orbital energies.

O,O'-diethyldithiophosphate complexes have been studied in P.E. spectroscopy independently by *Maier* and *Sweigart (101)*, *Fragalà et al. (102)*, and *Costanzo et al. (103)*. The P.E. spectra of the ligand in acid form and of the Cr(III), Co(III), Ni(II), Pd(II), Pt(II), and Zn(II) complexes resemble those of the corresponding difluorodithiophosphates in several basic features. Thus, the first ligand-based ionizations at ~ 9.9–10.1, 10.5–11.1, and 10.5–12.0 eV in $M(S_2PF_2)_n$ are matched by bands around 8.4–9.0, 9.2–10.0, and 10.1–10.6 eV in $M[S_2P(OC_2H_5)_2]_n$ [e.g., 8.44, 9.20, and 10.15 eV in Crdtp$_3$ *(103)*, the shift by ~ 1.7 eV to lower I.E. in the ethoxyderivative being due to the smaller $-$ I effect of $-OC_2H_5$ as compared with $-$ F. At higher I.E.s, dtp complexes exhibit additional features, including ionization of oxygen lone pairs and of the σ-electrons of the ethyl groups; the ionization patterns beyond ~ 11 eV are then more complex and less resolved and allow less clear and less complete assignments than for $S_2PF_2^-$ complexes. The structure of the extra bands which appear in the region of low I.E. is, however, parallel to that observed in $S_2PF_2^-$ complexes, e.g., there is a band at 7.56 in Crdtp$_3$ which can only be due to ionization of the d^3 configuration of Cr(III) *(103)*. For Nidtp$_2$ conflicting assignments were proposed: according to *Maier* and *Sweigart (101)* both the 7.24 and the 8.37 eV bands are to be assigned to d-ionizations, while *Fragalà* and coworkers *(102–103)* suggest both bands to be mainly ligand, not d, in character. It should be recalled that these bands correspond to the bonds at 8.76 and 9.84 eV in $Ni(S_2PF_2)_2$ which, according to *Andreocci (105)*, are identified as due to a sum of d-ionizations the first and to π-coordinative bonding orbitals the second. Such controversy illustrates the situation of uncertainty existing in the assignment of d^8 and more generally of late transition elements complexes. The assignment proposed by *Maier* and *Sweigart (101)* finds apparent support in the difference between the I.E. of the first two bands of $Ni(dtp)_2$ (~ 1.3 eV), which is close to a difference in d-orbital excitation energies found from single-crystal absorption spectroscopic studies of the same compound *(107)*; however, possible deviations from Koopmans' behavior, contributions from relaxation energies to I.E.s and deviations from interelectronic repulsion energies to the observed transition energies suggest that no strict correspondence is to be expected between P.E. ionization energies and absorption spectroscopic transition energies, and that the coincidence observed in the present case may be fortuitous.

The P.E. spectra of transition metal dialkyldithiocarbamates have some common features with dithiophosphate complexes but are usually less well resolved, hence less informative. The P.E. investigation of $(CH_3)_2NCS_2CH_3$ by *Guimon et al. (104)* indicates the energy region between ~ 8 and 10 eV as the region of sulfur π and sulfur lone pair ionizations, which appear slightly shifted, as usual, to lower energies in the P.E. spectra of the bisdiethyldithiocarbamates of Ni(II), Cu(II), and Zn(II) and in the tris-diethyldithiocarbamates of Cr(III), Fe(III) and Co(III) (~ 7.5–9.5 eV) *(108)*. The assumed order of orbital energies in this region is $\pi_3(S) > n_-(\sigma_{MS}) > \pi_2(S) > n_+(\sigma_{MS})$, but the bands are overlapping and only partially resolved, so as not to warrant specific one-to-one assignments. Smaller extra bands in the region of lower I.E. are supposed to be due to d^n ionizations, e.g., d^3 of Cr(III) at 7.02, and the first component of d^6 of Co(III) at 6.67 eV. The first broad band of Fedtc$_3$ does not give evidence for separate, early ionization of the d^5 system of Fe(III), as is also the case with the Cu(II) and Zn(II) complexes, while a comparison with the P.E. spectrum of Nidtc$_2$, having an additional, well-separated, multiple band as low as 6.95–7.13 eV, suggests that here too the d^8 system of Ni(II) is actually ionized before ligand-based coordination orbitals (*108*). Several substituted iron(II) carbonyl dithiocarbamates Fe(CO)$_2$(RR'NCSS)$_2$ have been examined in P.E. spectroscopy (*109*); these complexes lose readily CO on sublimation, and the observed P.E. spectra are those of the related Fe^{II}(RR'NCSS)$_2$ species, where no separate d-ionization is apparent, and the ligand bands are grouped together under a few broad and only partly resolved envelopes. The interesting point is that I.E. values depend strongly on the nature of the R and R' groups, spanning an interval of about 1 eV between the bis-phenyl derivative, which is the most easily ionized one in the series, and the dialkylsubstituted complexes.

Other complexes of anionic sulfur ligands include the thioderivatives of β-dike tonates discussed in the preceding paragraph (*97, 98*).

Some Remarks and Comments

This review has no final conclusions. The whole matter of P.E. spectra of volatile metal compounds is still under investigation; the results obtained until now yield only a partial picture, and there are several fundamental problems still open, so generalized conclusions are not warranted at the present stage of research. However, some points regarding the significance of the P.E. spectroscopic technique in coordination chemistry are already self-evident. We shall try to identify open problems, lines of future research, and precautions to be taken both in the experimental research and in the interpretive work, at least in the form and to the extent suggested by the present partial stage of the development of research in this field.

To the coordination chemist, one of the most interesting results expected from the gas phase P.E. technique is the identification of the energy levels of the valence shell mainly composed of d-orbitals, yielding information complementary to those supplied by electronic absorption spectroscopy, especially on systems inaccessible to direct absorption-spectroscopic investigation of the d-shell, e.g., d^{10} species. P.E. spectroscopy comes indeed very close to this goal and has proved capable of yielding a vast amount of information on the energies in the d-shell of complexes, many of which are not otherwise available, and represent extremely valuable new material for checking the validity of ligand field and other bond models. Thus, ligand field splitting patterns in d^{10} systems and/or complexes of zero- or low-valent metals come out neatly from P.E. spectra, e.g., of carbonyls, sandwich complexes, and halides. It turns out frequently that d-ionization energies are higher than those of coordinative bond orbitals or of ligand-based orbitals, thus suggesting that d-orbital energies may be lower than the energies of σ and π coordinative bonding electrons or of ligand orbitals, even when d-orbitals are the first ones to be excited at low photon energies in the course of absorption spectroscopic transitions. Such a situation is not unexpected and can be explained by the effect of contributions from electronic repulsion terms to spectroscopic transition energies, and from relaxation corrections to P.E. ionization energies. However, this situation can be only guessed at and receives no or only indirect evidence from absorption spectra, whereas it becomes primary evidence in P.E. ionization patterns.

Two facts oppose the full achievement, in a simple way, of the goal of exact knowledge of orbital energy levels from P.E. spectra. The first, experimental in nature, is the difficulty of identifying the P.E. spectral bands actually due to d-ionizations; the second is interpretive, viz. the imperfect correspondence between orbital and ionization energies, and our insufficient knowledge of how to compute corrections, or more generally the mechanism of deviations from Koopmans' behavior. As to the latter, the obvious implications are a warning not to attach uncritically the meaning of sequence of orbital energies to the sequence of experimental P.E. ionization energies (a confusion which results more dangerous with d-orbital systems than with molecules containing only light or nontransitional atoms); more theoretical work is needed to elucidate the deviations from Koopmans' behavior and to search for rapid, efficient and more generally valid computational schemes for prediction of ionization energies.

The point of the identification of d-ionization bands in complex P.E. spectral patterns of coordination compounds lends itself to several empirical considerations. From the analysis of available data and within the limitations and precautions mentioned above, we can infer that the actual values of I.E. for d-orbitals (strictly for orbitals extending over the whole molecule of a coordination compound and mainly composed of the metal-centered d-orbitals) depend on:

i. The atomic number and the effective oxidation state of the central metal atom. Early transition metals (e.g., Ti and V) have high d-orbital energies and relatively unshielded d-electron distributions, so that changes in the environmental potential induce large changes in orbital energies, hence also in ionization energies. For example, V(IV) d^1 is ionized at relatively low I.E.s, which range from 6.2 eV in $V[N(CH_3)_2]_4$ (24) to 9.41 eV in VCl_4 (19). Late transition elements (e.g., Ni and Cu) have not only lower, but also more constant d-ionization energies, whose values are less sensitive to environmental changes. It may prove important for assignment purposes that the region where the ionizations of the d^8 electron system of Ni(II) in quadratic complexes are observed or suspected is typically comprised in all known cases between ~ 7 and 9 eV, even including the spread of ligand-field splitting and large variations in inductive effects of the ligands. Analogously, d-ionizations in Cu(II) complexes can be guessed at in a region limited to 7.6–9.7 eV and are actually observed in Hg(II) linear species between ~ 15 and 19.6 eV.

ii. Inductive effects from the ligands. Typical and well characterized is the effect of fluorine substituents, which raise, for example, all ionization energies of metal β-diketonate complexes by about 1.0 eV for each replacement of a $-CH_3$ through a $-CF_3$ group in the series acac–tfac–hfac, and the I.E.s of $F_2PS_2^-$ complexes by an amount of the same order with regard to those of the analogous complexes of $(C_2H_5O)_2PS_2^-$. Such effects tend to shift parallely the whole P.E. spectrum, including d- as well as ligand ionizations, and do not affect strongly the relative energy position of metal and ligand orbitals. Quite different is the effect of changing the donor atoms, e.g., between β-diketones and their thioderivatives, or their Schiff bases, which affects primarily the energy of the donor orbitals and may change, sometimes even drastically, the relative position of ligand and metal orbitals.

iii. The availability of mechanisms by which metal d-orbitals can become delocalized onto the ligands and become mixed with their orbitals, losing part of their d-character. We mention here again the comparison between complexes of the $-N(CH_3)_2$ and $-N[Si(CH_3)_3]_2$ ligands (24, 25, 92), where $N \rightarrow Si$ π-bonding enhances substantially the delocalization of metal d-electrons on the nitrogen ligands, so as to make their I.E. higher than the ionization energy of nitrogen lone pairs, contrary to what happens in the $-N(CH_3)_2$ complexes.

Whether or not the d-ionization bands appear separately before the ligand ionization bands depends essentially on the relative position in the energy scale of the d- and ligand orbitals contributing to the molecular structure of the complexes, subject to effects i, ii, and iii. This leads in general to occurrence of d-ionization bands before any other orbital ionization in complexes of early transition metals having few d-electrons in their partly filled shell, e.g., d^1 of Ti(III), d^2 of V(III) or d^3 of Cr(III), while d^9 and d^{10} have never been observed to give rise to early ionizations in compounds of metals in positive oxidation states; in intermediate cases, the most common of which is the d^8 configuration of Ni(II) complexes, both situations may occur; we are inclined to assume that d-ionizations occur earlier whenever the ligands are highly electronegative, e.g., $F_2PS_2^-$, or fall definitely below ligand ionizations whenever the latter bear highly negative charges on the donor atoms, or for any other reason, e.g., π-conjugation, have high orbital energies (porphyrins, Schiff bases), whereas mixed and not easily identifiable patterns occur in intermediate cases. The whole matter is, however, to be regarded as unsettled and controversial.

It should also be recalled that there are empirical checks which can be used for identification purposes: first of all the increase in intensity of d-ionization bands often observed when He(II) is used instead of He(I) as the source of ionization. A more widespread use of such experimental aid is to be strongly recommended and will probably be one of the most fruitful experimental developments in this field.

The observation of d-orbitals is not the only information obtainable from UPS spectra, nor probably the most important, since a P.E. spectrum is a display showing at the same level of evidence and detail the system of *all* the energy levels corresponding to all molecular orbitals of the valence shell of a volatile coordination compound, the d-based as well as the ligand-based orbitals. It has therefore the merit of showing with utmost evidence the arbitrariness of focussing the investigation on the d-shell only, in what we might call a sort of professional deformation more or less consciously suggested or promoted by the uncritical use of the methods and principles of ligand-field theory.

There are further possible advantages from the use of P.E. spectroscopy in the elucidation of the electronic structure of coordination compounds, and information not otherwise available or only indirectly available may thus come to light. The energy position of ionization bands of nonbonding orbitals, such as $6\,s^2$ in (Tl(I), $5\,s^2$ in Sn(II), or $5\,d\delta$ in Hg(II) compounds, can be simply related to the net overall atomic charge, within the limits to which ionization energies can be assumed to reproduce orbital energy sequences. It is hardly necessary to underline the importance of such a simple and straightforward determination of a fundamental quantity which is still not amenable to direct experimental measurement, and is subject to controversial estimates.

List of Abbreviations and Symbols

acac	Acetylacetonate	mes	Mesitylene
BAE	N,N'-Ethylene-bis-(acetylacetoneimi-nate)	M.O.	Molecular orbital
		P.E.	Photoelectron
bz	Benzene	Sacac	Monothioacetylacetonate
ced	Cycloheptadienyl	SacSac	Dithioacetylacetonate
cet	Cycloheptatrienyl	Stfac	Monothiotrifluoroacetylacetonate
chd	Cyclohexadienyl	S_2tfac	Dithiotrifluoroacetylacetonate
cot	Cyclooctatetraenyl	tfac	Trifluoroacetylacetonate
cp	Cyclopentadienyl	tol	Toluene
dtc	N,N'-Diethyldithiocarbamate	TPP	Tetraphenylporphine
dtp	0,0'-Diethyldithiophosphate	UPS	Ultraviolet photoelectron spectroscopy
Et	Ethyl		
hfac	Hexafluoroacetylacetonate	γ_i, Γ_i	Irreducible representation, denoted in the text and in the tables by small letters if referring to M.O.s of the parent molecule and by capital letters if referring to many-electron states of the +1 ions.
I.E.	Ionization energy		
I.P.	Ionization Potential		
K.E.	Kinetic energy		
Me	Methyl		
Mecp	Methylcyclopentadienyl		

References

1. *Hamnett, A., Orchard, A. F.:* Electronic Structure and Magnetism of Inorganic Compounds. London: Chem. Soc. 1972, Vol. I, p. 1.
2. *James, T. L.:* J. Chem. Education **48**, 712 (1971).
3. *Baker, A. D., Brundle, C. R., Thompson, M.:* Chem. Soc. Rev. **3**, 355 (1972).
4. *Orchard, A. F.:* Electronic States of Inorganic Compounds: New Experimental Techniques. Day, P. (ed.) p. 267. Dordrecht: D. Reidel 1975.
5. *Koopmans, T.:* Physica **1**, 104 (1934).
6. *Turner, D. W., Baker, C., Baker, A. D., Brundle, C. R.:* Molecular Photoelectron Spectroscopy. London: Wiley Interscience 1970, p. 264.
7. *Evans, S., Hamnett, A., Orchard, A. F.:* Chem. Comm. 1282 (1970)
8. *Lloyd, D. R.:* Chem. Comm. 868 (1970)
9. *Evans, S., Hamnett, A., Orchard, A. F.:* J. Coord. Chem. **2**, 57 (1972).
10. *Rohmer, M., Veillard, A.:* Chem. Comm. 250 (1973).
11. *Hillier, I. H., Guest, M. F., Higginson, B. R., Lloyd, D. R.:* Mol. Phys. **27**, 215 (1974).
12. *Guest, M. F., Higginson, B. R., Lloyd, D. R., Hillier, I. H.:* J. Chem. Soc. [Faraday II] **71**, 902 (1975).
13. *Berkowitz, J.:* J. Chem. Phys. **61**, 407 (1974).
14. *Connor, J. A., Derrick, L. M. R., Hall, M. B., Hillier, I. H., Guest, M. F., Higginson, B. R., Lloyd, D. R.:* Molec. Phys. **28** (5), 1193 (1974).
15. *Higginson, B. R., Lloyd, D. R., Evans, S., Orchard, A. F.:* J. Chem. Soc. [Faraday II] **71**, 1913 (1975).
16. *Guest, M. F., Hillier, I. H., Higginson, B. R., Lloyd, D. R.:* Molec. Phys. **29** (1), 113 (1975).
17. *Hall, M. B., Hillier, I. H., Connor, J. A., Guest, M. F., Lloyd, D. R.:* Molec. Phys. **30** (3), 839 (1975).
18. *Connor, J. A., Derrick, L. M. R.: Hillier, I. H., Guest, M. F., Lloyd, D. R.:* Molec. Phys. **31**, 23 (1976).
19. *Cox, P. A., Evans, S., Hamnett, A., Orchard, A. F.:* Chem. Phys. Lett. **7**, 414 (1970).
20. *Lloyd, D. R., Schlag, E. W.:* Inorg. Chem. **8**, 2544 (1969).
21. *Evans, S., Green, J. C., Orchard, A. F., Saito, T., Turner, D. W.:* Chem. Phys. Lett. **4**, 361 (1969).
22. *Higginson, B. R., Lloyd, D. R., Burroughs, P., Gibson, D. M., Orchard, A. F.:* J. Chem. Soc. [Faraday II] **69**, 1659 (1973).
23. *Baerends, E. J., Oudshoorn, Ch., Oskam, A.:* J. Electr. Spectr. **6**, 259 (1975).
24. *Gibbins, S. G., Lappert, M. F., Pedley, J. B., Sharp, G. J.:* J. Chem. Soc. Dalton 72 (1975).
25. *Lappert, M. F., Pedley, J. B., Sharp, G. J., Bradley, D. C.:* J. Chem. Soc. Dalton 1737 (1976).
26. *Alyea, E. C., Bradley, D. C., Copperthwaite, R. G., Sales, K. D.:* J. Chem. Soc. Dalton 185 (1973).
27. ref. 6, p. 36.
28. ref. 6, p. 361.
29. *Eland, J. H. D.:* Photoelectron Spectroscopy. London: Butterworths 1974, p. 196.
30. *Green, J. C., King, D. I., Eland, J. H. D.:* Chem. Comm. 1121 (1970).
31. *Müller, J., Fenderl, K., Mertschenk, B.:* Chem. Ber. **104**, 700 (1971).
32. *Hillier, I. H., Saunders, V. R., Ware, M. J., Bassett, P. J.; Lloyd, D. R., Lynaugh, N.:* Chem. Comm. 1316 (1970).
33. *Head, R. A., Nixon, J. F., Sharp, G. J., Clark, R. J.:* J. Chem. Soc. Dalton 2054 (1975).
34. *Bassett, P. J., Higginson, B. R., Lloyd, D. R., Lynaugh, N., Roberts, P. J.:* J. Chem, Soc. Dalton 2316 (1974).
35. *Higginson, B. R., Lloyd, D. R., Connor, J. A., Hillier, I. H.:* Chem. Soc. [Faraday II] **70**, 1418 (1974).
36. *Hall, M. B.:* J. Am. Chem. Soc. **97**, 2057 (1975).

37. *Guest, M. F., Hall, M. B., Hillier, I. H.:* Mol. Phys. **25**, 629 (1973).
38. *Hall, M. B., Guest, M. F., Hillier, I. H.:* Chem. Phys. Lett. **15**, 592 (1972).
39. *Evans, S., Green, J. C., Orchard, A. F., Turner, D. W.:* Disc. Faraday Soc. **47**, 112 (1969).
40. *Lichtenberger, D. L., Sarapu, A. C., Fenske, R. F.:* Inorg. Chem. **12**, 702 (1973).
41. *Cradock, S., Ebsworth, E. A. V., Robertson, A.:* J. Chem. Soc. Dalton 22 (1973).
42. *Cradock, S., Ebsworth, E. A. V., Robertson, A.:* Chem. Phys. Lett. **30**, 413 (1975).
43. *Lichtenberger, D. L., Fenske, R. F.:* J. Am. Chem. Soc. **98**, 50 (1976).
44. *Hall, M. B., Fenske, R. F.:* Inorg. Chem. **11**, 768 (1972).
45. *Lichtenberger, D. L., Fenske, R. F.:* Inorg. Chem. **13**, 486 (1974).
46. *Dewar, M. J. S., Worley, S. D.:* J. Chem. Phys. **51**, 1672 (1969).
47. *Dewar, M. J. S., Worley, S. D.:* J. Chem. Phys. **50**, 654 (1969).
48. *Derrick, P. J., Åsbrink, L., Edqvist, O., Jonsson, B. Ö., Lindholm, E.:* Int. J. Mass. Spectr. Ion Phys. **6**, 203 (1971).
49. *Evans, S., Green, M. L. H., Jewitt, B., Orchard, A. F., Pygall, C. F.:* J. Chem. Soc. [Faraday II] **68**, 1847 (1972).
50. *Condorelli, G., Fragalà, I., Centineo, A., Tondello, E.:* J. Organomet. Chem. **87**, 311 (1975).
51. *Green, J. C., Jackson, S. E., Higginson, B.:* J. Chem. Soc. Dalton 403 (1975).
52. *Evans, S., Green, M. L. H., Jewitt, B., King, G. H., Orchard, A. F.:* J. Chem. Soc. [Faraday II] **70**, 356 (1974).
53. *Rabalais, J. W., Werme, L. D., Bergmark, T., Karlsson, L., Hussain, M., Siegbahn, K.:* J. Chem. Phys. **57**, 1185 (1972); erratum, ibid. **57**, 4508 (1972).
54. *Cox, P. A., Evans, S., Orchard, A. F.:* Chem. Phys. Lett. **13**, 386 (1972).
55. *Groenenboom, C. J., de Liefde Meijer, H. J., Jellinek, F., Oskam, A.:* J. Organomet. Chem. **97**, 73 (1975).
56. *Evans, S., Green, J. C., Jackson, S. E., Higginson, B.:* J. Chem. Soc. Dalton 304 (1974).
57. *Evans, S., Green, J. C., Jackson, S. E.:* J. Chem. Soc. [Faraday II] **68**, 249 (1972).
58. *Herring, F. G., McLean, R. A. N.:* Inorg. Chem. **11**, 1667 (1972).
59. *Evans, S., Orchard, A. F., Turner, D. W.:* Int. J. Mass Spectr. Ion. Phys. **7**, 261 (1971).
60. *Evans, S., Guest, M. F., Hillier, I. H., Orchard, A. F.:* J. Chem. Soc. [Faraday II] **70**, 417 (1974).
61. *Clark, J. P., Green, J. G.:* J. Organomet. Chem. **112**, C 14 (1976).
62. *Green, J. C., Jackson, S. E.:* J. Chem. Soc. Dalton 1698 (1976).
63. *Symon, D. A., Waddington, T. C.:* J. Chem. Soc. Dalton 2140 (1975).
64. *Whitesides, T. H., Lichtenberger, D. L., Budnik, R. A.:* Inorg. Chem. **14**, 68 (1975).
65. *Lichtenberger, D. L., Sellmann, D., Fenske, R. F.:* J. Organomet. Chem. **117**, 253 (1976).
66. *Lappert, M. F., Pedley, J. B., Sharp, G. J., Westwood, N. P. C.:* J. Electr. Spectr. **3**, 237 (1974).
67. *Barker, G. K., Lappert, M. F., Pedley, J. B., Sharp, G. J., Westwood, N. P. C.:* J. Chem. Soc. Dalton 1765 (1975).
68. *Dehmer, J. L., Berkowitz, J., Cusachs, L., Aldrich, H. S.:* J. Chem. Phys. **61**, 594 (1974).
69. *Green, J. C., Green, M. L. H., Joachim, P. J., Orchard, A. F., Turner, D. W.:* Philos. Trans. R. Soc. London [A] **268**, 111 (1970).
70. *Burroughs, P., Evans, S., Hamnett, A., Orchard, A. F., Richardson, N. V.:* J. Chem. Soc. [Faraday II] **70**, 1895 (1974).
71. *Diemann, E., Müller, A.:* Chem. Phys. Lett. **19**, 538 (1973).
72. *Forster, S., Felps, S., Cusachs, L. C., McGlynn, S. P.:* J. Am. Chem. Soc. **95**, 5521 (1973).
73. *Lee, T. H., Rabalais, J. W.:* Chem. Phys. Lett. **34**, 135 (1975).
74. *Vonbacho, P. S., Saltsburg, H., Ceasar, G. P.:* J. Electr. Spectr. **8**, 359 (1976).
75. *Cocksey, B. G., Eland, J. H. D., Danby, C. J.:* J. Chem. Soc. [Faraday II] **69**, 1558 (1973).
76. *Boggess, G. W., Allen, J. D., Jr., Schweitzer, G. K.:* J. Electr. Spectr. **2**, 467 (1973).
77. *Orchard, A. F., Richardson, N. V.:* J. Electr. Spectr. **6**, 61 (1975).
78. *Eland, J. H. D.:* Int. J. Mass Spectr. Ion. Phys. **4**, 37 (1970).
79. *Burroughs, P., Evans, S., Hamnett, A., Orchard, A. F., Richardson, N. V.:* Chem. Comm. 921 (1974).

80. *Wittel, K., Bock, H., Haas, A., Pflegler, K. H.:* J. Electr. Spectr. **7**, 365 (1975).
81. *Schmidt, H., Schweig, A., Manuel, G.:* J. Organomet. Chem. **55**, C 1 (1973).
82. *Schmidt, H., Schweig, A., Manuel, G.:* Chem. Comm. 667 (1975).
83. *Berkowitz, J.:* J. Chem. Phys. **56**, 2766 (1972).
84. *Dehmer, J. L., Berkowitz, J., Cusachs, L. C.:* J. Chem. Phys. **58**, 5681 (1973).
85. *Williams, D. R., Poole, R. T., Jenkin, J. G., Liesegang, J., Leckey, R. C. G.:* J. Electr. Spectr. **9**, 11 (1976).
86. *Streets, D. G., Berkowitz, J.:* Chem. Phys. Lett. **38**, 475 (1976).
87. *Evans, S., Orchard, A. F.:* J. Electr. Spectr. **6**, 207 (1975).
88. *Lappert, M. F., Pedley, J. B., Sharp, G.J.:* J. Organomet. Chem. **66**, 271 (1974).
89. *Evans, S., Green, J. C., Jackson, S. E.:* J. Chem. Soc. [Faraday II] **69**, 191 (1973).
90. *Galyer, L., Wilkinson, G., Lloyd, D. R.:* Chem. Comm. 497 (1975).
91. *Cradock, S., Savage, W.:* Inorg. Nucl. Chem. Lett. **8**, 753 (1972).
92. *Harris, D. H., Lappert, M. F., Pedley, J. B., Sharp, G. J.:* J. Chem. Soc. Dalton 945 (1976).
93. *Starzewski, K. A. O., Dieck, H. T., Bock, H.:* J. Organomet. Chem. **65**, 311 (1974).
94. *Khandelwal, S. C., Roebber, J. L.:* Chem. Phys. Lett. **34**, 355 (1975).
95. *Evans, S., Hamnett, A., Orchard, A. F., Lloyd, D. R.:* Disc. Faraday Soc. **54**, 227 (1972).
96. *Brittain, H. G., Disch, R. L.;* J. Electr. Spectr. **7**, 475 (1976).
97. *Cauletti, C., Furlani, C., Storto, G.:* (1977) (to be published).
98. *Cauletti, C., Furlani, C.:* J. Electr. Spectr. **6**, 465 (1975).
99. *Condorelli, G., Fragalà, I., Centineo, G., Tondello, E.:* Inorg. Chim. Acta **7**, 725 (1973).
100. *Cauletti, C., Furlani, C., Piancastelli, M. N.:* J. Microsc. Spectr. Electron. **1**, 463 (1976).
101. *Maier, J. P., Sweigart, D. A.:* Inorg. Chem. **15**, 1989 (1976).
102. *Fragalà, I., Giuffrida, S., Ciliberto, E., Condorelli, G.:* Chim. Ind. (Milan) **58**, 656 (1976).
103. *Fragalà, I., Giuffrida, S., Ciliberto, E., Granozzi, G., Aiò, D.:* Proc. IX Nat. Conf. Inorg. Chem. (A.I.C.I.), 1976, paper F 10.
104. *Guimon, C., Goubeau, D., Pfister-Guillonzo, G., Åsbrink, L., Sansström, J.:* J. Electr. Spectr. **4**, 49 (1974).
105. *Andreocci, M. V., Dragoni, P., Flamini, A., Furlani, C.:* Inorg. Chem. **17**, 291 (1978).
106. *Bassett, R. J., Lloyd, D. R.:* J. Chem. Soc. Dalton 248,(1972).
107. *Tomlinson, A. A. G., Furlani, C.:* Inorg. Chim. Acta **3**, 487 (1969).
108. *Cauletti, C., Furlani, C.:* J. Chem. Soc. Dalton 1068 (1977).
109. *Cauletti, C., Duffy, N. V., Furlani, C.:* Inorg. Chim. Acta **23**, 181 (1977).
110. *Caesar, G. P., Milazzo, P., Cihonski, J. L., Levenson, R. A.:* Inorg. Chem. **13**, 3035 (1974).
111. *Nixon, J. F.:* J. Chem. Soc. Dalton, 2226 (1973).
112. *Berkowitz, J.:* Electron Spectroscopy. Shirley, D. A. (ed.). p. 391. Amsterdam: North-Holland 1972.
113. ref. 29, p. 20.
114. *Wittel, K., Mohanty, B. S., Manne, R.:* J. Electr. Spectr. **5**, 1115 (1974).
115. *Berkowitz, J., Dehmer, J. L.:* J. Chem. Phys. **57**, 3194 (1972).
116. *Evans, S.:* Disc. Faraday Soc. **54**, 143 (1972).
117. *Lloyd, D. R.:* Electron Spectroscopy. Shirley, D. A. (ed.). p. 445. Amsterdam: North-Holland 1972.
118. *Green, J. C., Powell, P., van Tilborg, J.:* J. Chem. Soc. Dalton 1974 (1976).
119. *Green, J. C., Hayes, A. J.:* Chem. Phys. Letters **31**, 306 (1975).
120. *Garner, C. D., Hillier, I. H., Guest, M. F., Green, J. C., Coleman, A. W.:* Chem. Phys. Letters **41**, 91 (1976).

169

Structure and Bonding

Editors: *J.D. Dunitz, P. Hemmerich, R.H. Holm,*
J.A. Ibers, C.K. Jørgensen, J.B. Neilands,
D. Reinen, R.J.P. Williams

Springer-Verlag
Berlin
Heidelberg
New York

Springer-Verlag
Berlin
Heidelberg
New York